BUILDING FUTURE SECURITY

Strategies for Restructuring the Defense Technology and Industrial Base

CONGRESS OF THE UNITED STATES
OFFICE OF TECHNOLOGY ASSESSMENT

Recommended Citation:

U.S. Congress, Office of Technology Assessment, *Building Future Security,* OTA-ISC-530 (Washington, DC: U.S. Government Printing Office, June 1992).

For sale by the U.S. Government Printing Office
Superintendent of Documents, Mail Stop: SSOP, Washington, DC 20402-9328
ISBN 0-16-037975-X

Foreword

The collapse of the Soviet military threat holds out the prospect of a "peace dividend" in the form of a smaller and less costly defense establishment. But despite the end of the cold war, the United States still faces existing and emerging security threats, including the rise of regional powers, the proliferation of advanced conventional military technologies and weapons of mass destruction, and the possibility of a renewed global military threat in the distant future. The Nation will therefore continue to need a robust defense technology and industrial base (DTIB) that can develop, produce, and support appropriate military systems in peacetime and respond to additional military requirements in crisis or war.

Building Future Security, the final report of OTA's assessment of the U.S. defense technology and industrial base (DTIB), discusses strategies for moving to a smaller and more efficient DTIB over the next decade and maintaining that base in the future. It complements OTA's earlier report, *Redesigning Defense*, which developed a framework for analysis of future defense needs, postulated some desirable characteristics of the future DTIB, and outlined some broad strategic choices that will affect the future base. This framework provided the starting point for the current report, which assesses some specific policy options for restructuring the DTIB.

The principal finding of *Building Future Security* is that while powerful bureaucratic, economic, and political interests favor a proportional downsizing of the DTIB in which a maximum number of current firms or organizations would survive (albeit smaller and perhaps weaker), this approach would not best serve the Nation's defense needs. Instead, if these needs are to be met, *the anticipated cuts in defense spending will require a fundamental restructuring of the DTIB to 1) reallocate resources from short-term military capabilities to long-term military potential, and 2) exploit the synergies that can result from a closer integration of the R&D, production, and maintenance elements of the base.*

For example, the future DTIB might seek to integrate R&D and production through a "prototyping-plus" strategy that involves the continuous development and limited production of selected prototypes during the periods between full production programs. Defense manufacturing might be maintained through some combination of low-rate production, greater integration of the civil and military industrial bases, and changes in procurement of spare parts and maintenance services. It is clear that future managers of the DTIB will need a better understanding of all elements of the base and should seek to enhance the strength of the entire base rather than a single element.

In undertaking this assessment, OTA sought information and advice from a broad spectrum of knowledgeable individuals and organizations whose contributions are gratefully acknowledged. As with all OTA studies, the content of this report is the sole responsibility of the Office of Technology Assessment and does not necessarily represent the views of our advisers and reviewers.

John H Gibbons

JOHN H. GIBBONS
Director

Advisory Panel—Building Future Security: Strategies for Restructuring the U.S. Defense Technology and Industrial Base

Walter B. Slocombe, *Chair*
Caplin & Drysdale Chartered

Richard Bohlen
Consultant

Robert Calaway
President
Resources Management International, Inc.

Gordon Corlew
Vice President, Engineering and Production
AIL Systems, Inc.

Jacques S. Gansler
Senior Vice President
TASC

Julius Harwood
Consultant

William W. Kaufmann
Senior Fellow
The Brookings Institution

General P.X. Kelley
USMC (Ret.)

James L. Koontz
President & CEO
Kingsbury Machine Tool Corp.

Thomas L. McNaugher
Senior Fellow
The Brooking Institution

William McNeill
Professor Emeritus
Department of History
University of Chicago

John Mearsheimer
Chairman
Department of Political Science
University of Chicago

William J. Perry
Chairman & CEO
Technology, Strategies & Alliances

Donald W. Putnam
Corporate Director of Contracts
General Dynamics Corp.

Jack Ruina
Professor of Electrical Engineering
Center for International Studies
Massachusetts Institute of Technology

Howard D. Samuel
President
Industrial Union Department AFL-CIO

Wickham Skinner
Professor Emeritus
Business Administration
Harvard University

General William Y. Smith
USAF(Ret)

James Solberg
Professor
Engineering Research Center
Purdue University

Leonard Sullivan
Consultant
System Planning Corp.

Admiral Harry Train, USN, (Ret.)
Division Manager
Strategic Research and Management Services Division
Science Applications International Corp.

General John W. Vessey, Jr.
USA (Ret.)

Albert Wheelon
Consultant

NOTE: OTA appreciates and is grateful for the valuable assistance and thoughtful critiques provided by the advisory panel members. The panel does not, however, necessarily approve, disapprove, or endorse this report. OTA assumes full responsibility for the report and the accuracy of its contents.

OTA Project Staff—Building Future Security: Strategies for Restructuring The U.S. Defense Technology and Industrial Base

Lionel S. Johns, *Assistant Director, OTA*
Energy, Materials, and International Security Division

Alan Shaw, *International Security and Commerce Program Manager*

Jack H. Nunn, *Project Director*

Ivan Oelrich

Jonathan B. Tucker

Christopher Waychoff

Administrative Staff

Jacqueline Robinson-Boykin, *Office Administrator*

Louise Staley, *Administrative Secretary*

Contractors

Madeline Gross

Ivars Gutmanis

John Jury

Deborah Shapley

Peter Tarpgaard

Contents

Chapter 1

Summary and Conclusions

Contents

Boxes

Figures

Tables

Chapter 1
Summary and Conclusions

INTRODUCTION

The transformation of the global security environment is causing sweeping changes in the U.S. defense technology and industrial base (DTIB). The collapse of the Soviet military threat, which drove U.S. defense planning and spending for 40 years, combined with the urgency of domestic problems and the spiraling budget deficit, have generated pressures to reduce the defense budget by a third to a half over the next decade. Yet the Persian Gulf War illustrated the continuing need for an effective U.S. military establishment, supported by a smaller but still robust DTIB.

Cuts in funding for defense research, development, production, and maintenance could impair the ability of the base to meet future national security needs unless the cuts are accompanied by changes in how the base is structured. As a result, the Nation needs to develop a comprehensive strategy for managing the downsizing of the DTIB while preserving the core capabilities essential for the development, production, and maintenance of major weapons and defense equipment. The broad outline of such a strategy was examined in an earlier OTA report, *Redesigning Defense* (See box 1-A.), and in three background papers.[1] The previous report described some desirable characteristics of the future DTIB, which are listed in table 1-1. This report elaborates on the findings of the earlier OTA publications and examines in greater detail the specific policy choices involved in restructuring the DTIB over the next decade.

Implications of Defense Budget Cuts

Both the administration and Congress appear to be preparing for major, long-term reductions in defense spending. The administration's fiscal year 1993 Department of Defense (DoD) budget request is for $267.6 billion in budget authority and $272.8 billion in outlays—a 7-percent reduction after inflation from the fiscal year 1992 spending level. The DoD projects that by 1997, budget authority will fall below $240 billion in constant 1992 dollars. (See table 1-2.)[2] Many members of Congress have proposed even deeper cuts.[3] By the end of the decade, the defense budget could well be between $180 and $220 billion in 1992 dollars. Even these projected cuts may be conservative given the continuing decline of the immediate military threat, the growing Federal budget deficit, and competing social priorities.[4]

Reductions in defense spending are likely to affect procurement accounts more than other budget areas. The fiscal year 1993 budget request, for example, put a cap on B-2 bomber production at 20, terminated the SSN-21 Seawolf attack submarine with the lead boat, shifted the focus of the Army's Comanche helicopter program from full production to building prototypes and developing subsystems, and terminated or reduced a host of other weapons.[5] A recent Congressional Budget Office report concluded that future budget cuts would leave little room for new weapon programs in the near term.[6] Further, DoD funding for procurement is likely to be constrained by competing demands. For example, the House Armed Services Committee noted in its fiscal year 1990 authorization report that compli-

[1] U.S. Congress, Office of Technology Assessment, *Adjusting to a New Security Environment: The Defense Technology and Industrial Base Challenge*, OTA-BP-ISC-79 (Washington, DC: U.S. Government Printing Office, February 1991); *Redesigning Defense: Planning the Transition to the Future U.S. Defense Industrial Base*, OTA-ISC-500 (Washington, DC: U.S. Government Printing Office, July 1991); *American Military Power: Future Needs, Future Choices*, OTA-BP-ISC-80 (Washington, DC: U.S. Government Printing Office, October 1991); *Lessons in Restructuring Defense Industry: The French Experience*, OTA-BP-ISC-96 (Washington, DC: U.S. Government Printing Office, June 1992).

[2] Secretary of Defense Dick Cheney, *Report of the Secretary of Defense to the President and Congress* (Washington, DC: U.S. Government Printing Office, February 1992), p. 21.

[3] Eric Schmitt, "Move to Shift $15 Billion from Military Gains Support in House," *New York Times*, p. A12, reports that Congressman Les Aspin has proposed a $91 billion cut from the budget by 1997.

[4] OTA's assessment of opportunities for economic conversion has recently been published in a report that addresses many of the problems of economic adjustment. U.S. Congress, Office of Technology Assessment, *After The Cold War: Living with Lower Defense Spending*, OTA-ITE-524 (Washington, DC: U.S. Government Printing Office, February 1992).

[5] Cheney, op. cit., footnote 2, p. 25. These proposals are all being hotly debated, particularly the termination of the Seawolf.

[6] Congressional Staff Memorandum, *Implications of Additional Reductions in Defense Spending*, Congressional Budget Office, October 1991.

Box 1-A— OTA's Redesigning Defense *Report*

Redesigning Defense described the defense technology and industrial base (DTIB) and current pressures to reduce it, and developed a framework for debating the size and structure of the future base. The report postulated some desirable characteristics of a future base, described the broad strategic choices that the Nation faces regarding the future base, and outlined tactical decisions that could be made to support the transition to the future base. The report's key findings are outlined below.

Definition of the DTIB—The defense technology and industrial base is defined as the combination of people, institutions, technological know-how, and facilities used to design, develop, manufacture, and maintain the weapons and supporting defense equipment needed to meet U.S. national security objectives.

The DTIB consists of three broad elements—research and development (R&D), production, and maintenance. Each of these has a private and a public component. The DTIB can also be divided into tiers—prime contractors, subcontractors, and parts and raw-material suppliers—and into different industrial sectors. While the DTIB is often discussed as if it were an independent entity, it is really interwoven with the Nation's civilian technology and industrial base and, increasingly, with the global economy.

Current Base Conditions—The report noted that although the DTIB has produced some outstanding weapons, as demonstrated in the Persian Gulf War, it has serious weaknesses that limit its ability to support future national defense needs for peacetime production and crisis response. Other studies have documented the problems of the high cost of weapon systems, growing dependence on foreign sources for critical components, and the shrinking number of defense subcontractors.

Desirable Characteristics of the Future Base—To avoid a weakened and potentially crippled DTIB, it is important to set goals for the future base. OTA suggested a list of desirable characteristics for a future DTIB as a guide for planning. These characteristics are outlined in table 1-1 of the text.

Broad Strategic Choices—The Nation needs a long-term strategy for identifying and maintaining critical facilities, technological know-how, and people needed to develop, manufacture, and maintain future systems. The Nation faces some broad strategic choices that will shape the future DTIB. Ad hoc decisions, made in lieu of a strategy, will result in a weak DTIB that will undermine the Nation's defense.

Autonomy v. Interdependence— The Nation must choose the degree of defense industrial autonomy that is necessary and possible in an increasingly global technological environment. There are risks both in excessive reliance on foreign sources and in attempting to be fully autonomous. In the former case, the Nation risks losing to offshore competitors critical capabilities and control over which technologies are pursued; in the latter case, it risks higher procurement costs, protected industries that lack innovative drive, and loss of access to foreign technological advances.

Arsenal System v. Civil Integration—A second choice relates the internal structure of the future base. On one hand, the Nation can rely on "arsenals," i.e., government or privately-owned, sole-source producers of particular military systems. On the other hand, the Nation can modify military requirements to allow much greater use of technologies in the civilian sector. In the absence of deliberate choices, the DTIB is likely to evolve towards an arsenal structure, since current procurement laws impede civil-military integration and shrinking production will lower the number of private defense contractors, thereby reducing competition.

Current Capability v. Future Potential—A third choice concerns the allocation of resources between current military capability and future military potential. Although some deployed capability is needed for future theater conflicts, the greatly reduced threat of a major global conflict allows a shift of funding away from production toward research and development.

Tactical Decisions—Besides the broad strategic choices mentioned above, the Nation needs to make tactical decisions to ensure that the future DTIB has the characteristics, outlined above, that are needed for a strong defense. These tactical decisions concern:

- Guiding and evaluating research and development
- Protecting core competencies
- Developing human resources
- Identifying critical manufacturing areas
- Setting manufacturing priorities
- Funding surge and mobilization planning

Redesigning Defense's description of desirable characteristics and the Nation's strategic and tactical decisions were the starting point for the current report. They were modified and extended as this second report developed.

Table 1-1—Desirable Characteristics of the Future Base

- Advanced research and development capability
- Ready access to civilian technology
- Continuous design and prototyping capability
- Limited, efficient peacetime engineering and production capabilities in key defense sectors
- Responsive production of ammunition, spares, and consumables for theater conflict
- Healthy, mobilizable civilian production capacity
- Robust maintenance and overhaul capability
- Good, integrated management

SOURCE: Office of Technology Assessment, 1991. Characteristics are not necessarily listed in order of priority.

Table 1-2—Department of Defense Budget Authority (billions of dollars)

Year	Current $	Constant $	Real growth %
1985	286.8	375.6	
1986	281.4	359.1	–4.4
1987	279.5	345.7	–3.8
1988	283.8	338.5	–2.1
1989	290.8	333.7	–1.4
1990[a]	291.0	324.1	–2.9
1991[a]	276.0	292.9	–9.6
1992[a]	277.5[b]	287.8[b]	–1.8
1993	267.6	267.6	–7.0
FY 1985-1993 real change:			**–28.8**
1994	267.8	258.0	–3.6
1995	269.9	250.4	–2.9
1996	270.4	241.8	–3.4
1997	274.6	237.5	–1.8
FY 1985-1997 real change:			**–36.8**

[a] Excludes cost of Operation DESERT SHIELD/STORM. This is consistent with the 1990 Budget Enforcement Act, which exempted DESERT SHIELD/STORM spending on an emergency basis from negotiated budget ceilings set by the Executive Branch and Congress. According to the DoD, the net U.S. cost for this operation should not exceed $5.9 billion after all foreign contributions are received.

[b] Enacted in FY 1992 DoD Appropriations Act. The FY 1992 figure in this year's budget request ($270.9 billion) differs because it reflects proposed environmental supplemental appropriations and proposed rescission of already appropriated funds.

SOURCE: Report of the Secretary of Defense to the President and Congress, February 1992.

ance with environmental legislation will cost the DoD $5 to $10 billion over the next 5 years.[7]

A recent DoD report on the defense industrial base noted, in something of an understatement, that "the consequences of DoD budget reductions will be one of the most important issues facing defense contractors in the 1990s."[8] Individual defense firms will need to restructure, and some face challenges to their survival. The government portion of the DTIB must also restructure as government-operated arsenals, depots, and laboratories are faced with the new national security and fiscal realities.

The DoD has asserted that its budget request reflects a new approach to defense acquisition, featuring:[9]

- heavy emphasis on government-sponsored R&D to maintain America's technology base;
- more reliance on prototyping, advancing to full production only after thorough testing and demonstration of a "critical" requirement;
- greater attention to the producibility of new systems and to manufacturing processes; and
- more reliance on upgrading and inserting new capabilities into existing platforms.

The DoD proposals embody many of the desirable DTIB characteristics discussed in *Redesigning Defense*. (See table 1-1 and box 1-A.) But while these policies represent the DoD's first real response to the challenges of the post cold war era, they are not sufficient to ensure an effective future base. **OTA's analysis indicates that a more detailed and integrated plan of action will be necessary if DoD initiatives are to result in a strong and healthy DTIB.** What is missing from the current approach is an announced strategy and an implementation plan (including budget considerations) that links these and other policies to ensure the ability of the DTIB to meet the Nation's future national security needs. Such an integrated approach is suggested later in this chapter.

ORGANIZATION OF THIS REPORT

This report consists of six chapters and one appendix. Using the desirable DTIB characteristics described in *Redesigning Defense* as a starting point, the chapters analyze detailed policies for achieving those characteristics. This chapter summarizes key findings and policy issues. Chapter 2 addresses alternatives for maintaining an advanced research and development (R&D) capability. Chapter 3 discusses OTA's "prototyping-plus" strategy and

[7] House Armed Services Committee, *FY 1990 Authorization Report.*

[8] Under Secretary of Defense (Acquisition) Assistant Secretary of Defense (Production and Logistics), *Report to Congress on the Defense Industrial Base,* November 1991, p. 4-1.

[9] Ibid.

its implications. Chapter 4 describes how a future production base might manufacture quality military equipment at an affordable price in peacetime and also meet the surge and mobilization requirements of a future crisis or war. Chapter 5 discusses policy alternatives for ensuring a robust maintenance capability. Chapter 6 considers the management of a future restructured base. The appendix summarizes plans of selected allied nations to deal with changes in their defense industrial bases.

The findings of the assessment are divided into general observations that apply to most or all of the DTIB, and more specific findings relating to one or a few parts of the base. The general findings also include a discussion of three issues that cut across all the elements of the base.

Photo credit: DoD

The reduced security threat of the near future can be met largely with existing inventories of weapons.

GENERAL FINDINGS

The capacity of the current U.S. DTIB to provide defense goods and services exceeds foreseeable national security requirements. This overcapacity is largely a result of the reduced military threat and the large inventory of military materiel on hand. However, the current base has potential production bottlenecks and shortfalls to quantity production that will be exacerbated as some producers are forced out of the defense business by cuts in funding. Reductions in capacity must therefore be undertaken with care.

Powerful military, economic, and political interests support downsizing the DTIB in a manner that allows the maximum number of current firms and organizations to survive, albeit reduced in size. Such a "proportional downsizing" would not best support the Nation's future defense needs. What is required is not just a smaller DTIB, but a restructured base with a new allocation of resources among its three main elements—R&D, production, and maintenance. The waning major military threat and large inventories of advanced weapons and equipment demand a relative shift of resources toward R&D, as has begun in recent defense budgets. The production and maintenance bases, while still important, will bear proportionally larger budget reductions.

The elements of the future DTIB must be better integrated. There must also be an integrated management approach that aims to achieve the best use of resources for the DTIB as a whole. In the past, DTIB managers have focused on achieving individual goals within their own organizations, with little attention given to the effects of these policies on the entire base. For example, R&D costs were made to appear artificially low by shifting some of the true cost of R&D to production. Government managers have also sought to control production costs through "spare parts breakout" (i.e., contracting production of spare parts to a firm other than the original equipment manufacturer) and the use of second sources. These policies, however, have reduced the funds available for full-service contractors to maintain R&D teams and facilities.

If the DTIB is to provide high-quality weapons at an affordable price in peacetime and to respond with increased production in crisis or war, it must be restructured to exploit the synergies arising from a closer integration of R&D, production, and maintenance. For example, R&D can be directed more toward improving production processes, and contractors can manufacture multiple products on a single production line that also upgrades older equipment. While rigid centralization is not an appropriate way to manage the future DTIB, it will be essential to develop an integrated management approach that gives priority to the needs of the entire base over those of its parts. Such an integrated approach may require reorganizing DoD oversight at the levels of the Office of the Secretary of Defense (OSD) and the individual Services to ensure an integrated approach to managing R&D, production, and maintenance. Managers at all levels may also need incentives to take a broader view of the base. R&D managers, for example, might be evaluated in part on their ability to promote the development of systems that can be manufactured more easily.

There may be a similar need to reorganize Congress' committee structure to improve communication among the various committees monitoring defense R&D, production, and maintenance. **The structural changes in the DTIB described in this report will require a concomitant shift in thinking about what constitutes national security and the role of science and industry in maintaining it. This new paradigm will rest on a willingness to purchase knowledge rather than hardware in many cases.** While standing forces are the currency of national power in a hot or cold war, military potential in the form of economic and technological strength is more important during periods of reduced military threats. Just as the Nation commits resources in peacetime to maintain divisions, air wings, and carrier battle groups against future contingencies, it must commit resources to preserve a strong DTIB.

The "pipeline" model of acquisition, which shapes current procurement policy and focuses on products rather than manufacturing processes, will be counterproductive in the future. Instead, flexibility in development and manufacturing will be essential. The automatic link between development and production could be broken so that the basic criterion for management success is not to produce a new system against all odds, but to develop new capabilities that only sometimes take the form of new hardware.

The current debate over maintaining a warm production base is incorrectly framed; the real issue is how to maintain a "warm capability." Such a capability can provide for the future development and production of new systems. Changing the terms of the debate in this way would provide the opportunity to identify the defense industrial sectors in which R&D alone is sufficient, those in which warm production lines must be preserved, and those in which other alternatives may exist.

Legislative and regulatory barriers impede civil-military integration. Current laws on defense acquisition aim to give a maximum number of companies access to public funds, while also ensuring maximum public accountability in the use of those funds. A negative effect of this approach has been to impose different regulatory and accounting rules for civil and defense activities, forcing firms to isolate their defense work from their civilian work. As the DTIB shrinks, this approach might be reexamined. Critics argue that greater integration between civil and military production would actually improve access by increasing the number of firms willing to do business with the DoD. This increase would in turn provide greater opportunities for competition and reduce the need for extraordinary government actions to ensure accountability. Although several DoD programs have sought to transfer more oversight responsibility to defense firms, these programs have often failed because of a lack of long-term support from the DoD acquisition community.

Since the DoD is unlikely to be at the forefront of all defense-relevant product and process technologies, it should establish priorities for which technologies it wishes to pursue. Defense-relevant technologies are increasingly developed in the civil sector and by other countries. The DoD needs to track these developments and to take advantage of them.

Cross-Cutting Issues

Some issues confront policymakers with common challenges across the DTIB. Three of these crosscutting issues—*human resources, facilities,* and *technology*—will be key to whether the United States has a strong DTIB in the 21st century.

Human Resources

People are the single most important ingredient of the DTIB. They provide the knowledge to conceive of and build new systems, devise and improve manufacturing processes, and manage the base. To retain a healthy DTIB, the Nation must therefore retain high-quality technical and managerial personnel, encourage them to improve their skills, and attract new people. Even more important than individuals are teams with special know-how that is passed down over decades, such as aircraft design teams and missile production groups. The continuity of such teams is critical to technical advancement and the Nation's future military capabilities. Yet private companies and government organizations are slashing personnel and training programs to remain competitive or simply to survive economically.

The objectives of a future DTIB human-resources policy should not be to retain the

maximum number of people currently employed in the defense industry but to ensure that individuals and teams with essential skills are preserved, and to help those who leave the DTIB to maintain relevant skills in the civil sector. The strategy for preserving skills (both individual and team) depends on the industrial sector. Electronics skills, for example, can be maintained in the civilian base with little government intervention. In contrast, since submarine production and munitions design are not performed in the civilian base, specific actions will be needed to preserve know-how in these areas.

Facilities

The cold war mobilized a significant portion of private industry and expanded the government's military research, production, and maintenance facilities. **The end of the cold war requires the demobilization of many private and government facilities.** Facilities can be replaced more easily than people, but some facilities are unique and not easily replicated once closed, including large dry docks, aerospace test facilities, special laboratories, and maintenance hangars. Nevertheless, the limitations on new production and maintenance work will make it costly for the Nation to maintain duplicate facilities. It will therefore be necessary to decide when to consolidate to a single facility, when to maintain more than one, when to rely on allied capabilities, and when to close a unique facility and adopt an entirely different approach to meeting a national security need.

The objective of a facilities strategy should not be to maintain current capacity, but to ensure the proper mix and size of future DTIB facilities. How this is done will vary by industrial sector and technology. The government may have to intervene to preserve militarily unique facilities for tank assembly, nuclear submarines, and ammunition. Technologies and industrial sectors with more civil applications (e.g., electronics, fasteners, and clothing) can probably be maintained entirely in the civil sector. Even so, this approach would require changes in DoD acquisition policy such as eliminating overly rigid military specifications and designing military systems to allow use of commercial components. Critical and unique facilities might be preserved either by allowing them to be used profitably in the private sector or by converting them into government-owned, contractor-operated (GOCO) facilities or government-owned, government-operated (GOGO) facilities. OTA's analysis suggests that many critical facilities can be maintained by encouraging greater civil-military integration or by concentrating activities at a few select facilities.

Photo credit: DoD

Production facilities built to meet the needs of the cold war are in excess of current requirements.

Technology

Advanced technology remains critical to the Nation's military strength. But the narrow focus on battlefield performance during the cold war should give way to a broader approach that takes account of defense manufacturing and maintenance issues and economic security. The cold war spurred an outpouring of U.S. technological innovations aimed at outperforming a quantitatively superior enemy on the battlefield and building a strategic nuclear deterrent. In the future, military innovation might be sustained with relatively less funding and reorganized to take advantage of scientific and technological advances in the U.S. civil sector and abroad. Policymakers will also need to identify technologies with the potential to solve national security problems (the aim of the congressionally mandated Critical Technology Plans) and make a long-term commitment to funding their development.

National Choices

In a general sense, the chief defense-management challenge of the next decade will be to maintain the U.S. advantage in defense-related technology and to produce high-quality military hardware on a much smaller defense budget. There are different ways of

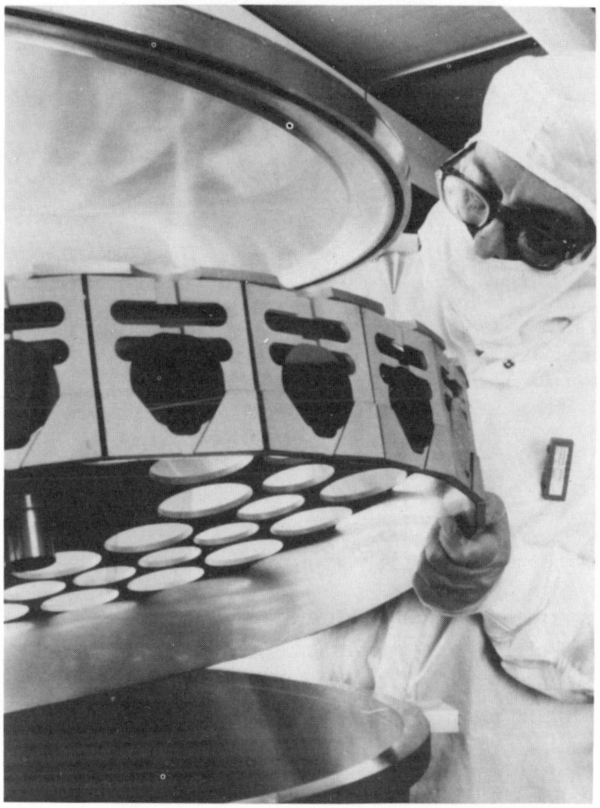

Photo credit: Lockheed/Sanders

A technician prepares gallium arsenide semiconductor wafers for a final manufacturing step. Such technologies will be crucial to future military capability.

organizing the future DTIB to achieve these goals. The alternative policies involve strategic choices, as described in the earlier OTA report *Redesigning Defense*. (See box 1-A.) Further analysis by OTA has led to refinement of these strategic choices, as indicated by the decision tree in figure 1-1. The strategic choices at each fork in the tree are not absolute but merely suggest general tendencies. For example, the United States might emphasize R&D and prototyping for most weapons systems, while still keeping some items in production at any given time to modernize selected portions of its forces.

The first choice for the Nation is between current and future military capabilities. To the extent the United States faces an immediate military threat, the DoD will need to allocate funds for current capabilities. If the immediate threat is reduced, however, the DoD has the option to shift funds to the development of military potential. The administration's fiscal year 1993 budget proposal made a tentative move in this direction by calling for a small real increase in R&D funding, a 13-percent decrease in procurement, and the cancellation of several production programs.

Subsequent choices are:

1. between dual-use and militarily unique technologies (both product and process);
2. between private and public ownership;
3. between competitive procurement and single sourcing; and
4. between reliance on domestic and international sources.

The decision tree outlines some of the reasons for making each of these choices.

For much of the military materiel required by the DoD, **OTA's analysis suggests that for reasons of cost, total capacity, and potential for innovation, the path defined by chosing dual-use technologies, private ownership, and competitive acquisition is preferable to alternate paths.** Nevertheless, in some cases other paths may be necessary because of unique military performance requirements or manufacturing processes (e.g., for production of ammunition and nuclear submarines), or technology security (e.g., for nuclear weapons).

Following the dual-use/private/competitive path would require a number of changes in current U.S. laws and regulations, including the adoption of accounting and manufacturing practices that do not isolate defense from civil production, a change in the profit/risk ratio for private-sector defense work, and an emphasis on flexible performance specifications rather than rigid military specifications for products and manufacturing processes.

Finally, there is a choice between national autonomy and international cooperation. OTA's analysis confirmed that this choice is important but is subordinate in most cases to other choices. In the new security environment, the government will need to ensure that the benefits of international arms collaboration, sales and purchases are weighed against the potential drawbacks.

SPECIFIC FINDINGS

The discussion below focuses on the desirable characteristics of the future DTIB (see table 1-1) and the policies for achieving them. It is important to keep in mind that these characteristics should be viewed as an integrated set. Policies developed for the R&D and maintenance elements, for example,

Figure 1-1—Strategic Choices for the Future DTIB

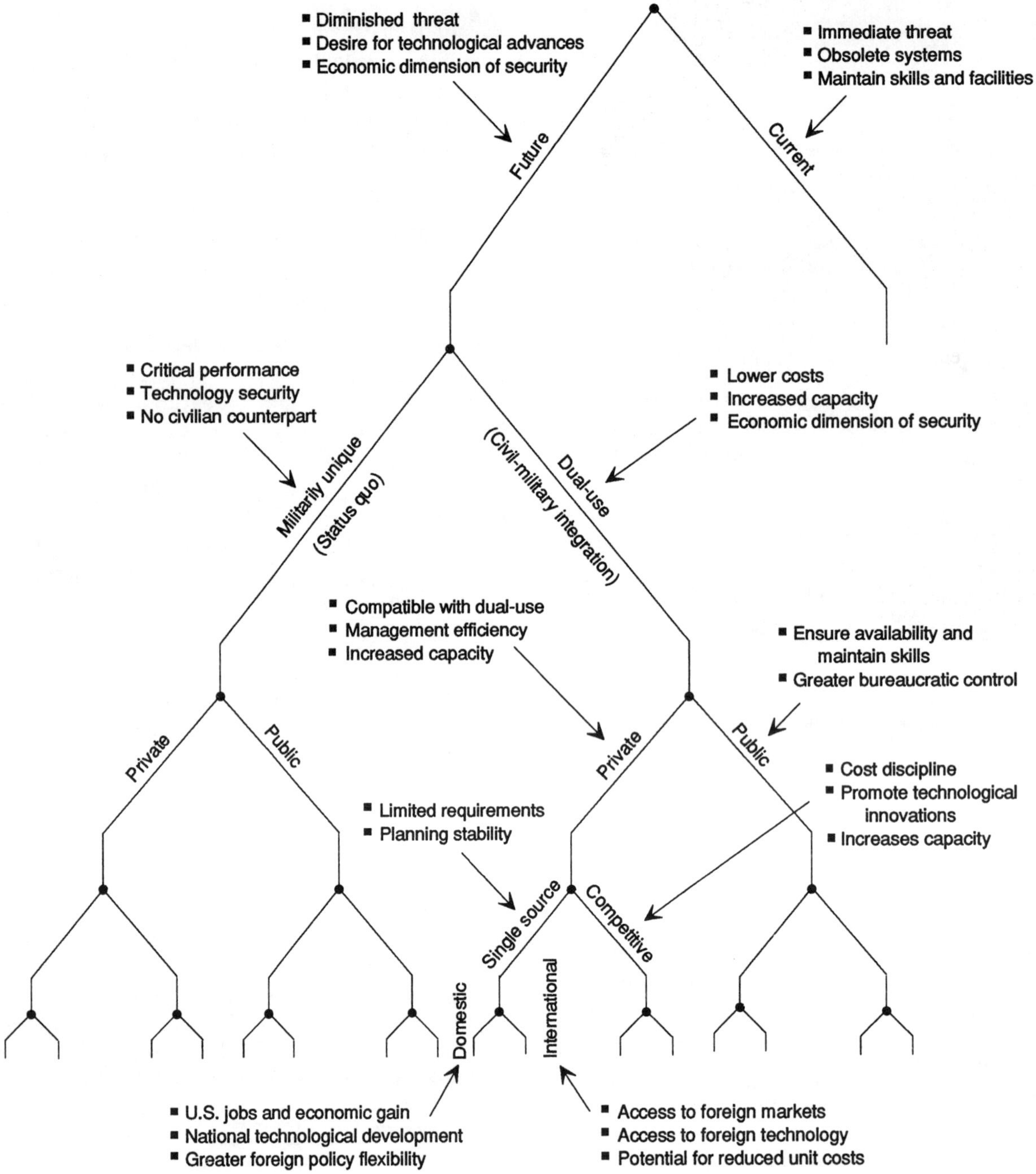

SOURCE: Office of Technology Assessment, 1992

will affect the health of the production base. A discussion of the integrated future base follows a description of the R&D, production, and maintenance elements.

Research and Development

Redesigning Defense stressed the importance to future national security of an advanced R&D capability that can 1) maintain qualitative weapon performance superiority against potential adversaries; 2) create opportunities for innovation and hedge against technological breakthroughs by opponents; and 3) support the Nation's overall economic strength, which is ultimately the source of its military strength.

An advanced defense R&D capability includes world-class personnel (individuals and teams); cutting-edge research that guards against technological surprise; robust efforts in critical technologies; a balance between the near-term technology needs of each Service and long-term U.S. defense needs; strong links to manufacturing, so that proposed weapon systems are producible; and integration with civilian R&D, even in the absence of a national consensus on directed federal support for civil technology programs.

Both the Administration and Congress have expressed a desire to support defense R&D. The DoD's current budget request contains a shift in relative emphasis toward R&D. Over the long term, however, the military R&D base will almost certainly shrink. Funding is expected to drop in real terms from around $40 billion today to $25 to $27 billion (in 1992 dollars) by 2001. Moreover, the DoD will have to pay explicitly for defense R&D rather than follow the past practice of funding it partially through production.

Without offsetting actions, funding reductions will result in disproportionate cuts in defense R&D performed by private industry. Direct R&D contracts to industry will decline, and lower procurement budgets will also reduce companies' willingness to invest their own money in R&D. DoD support for research in colleges and universities could also decline as the defense budget shrinks. As a result of these trends, the DoD will not have the benefits of some leading-edge research by industry and universities that it has enjoyed in the past. The DoD will also have less of a chance to familiarize the next generation of scientists and engineers with the Nation's defense needs. A national DTIB strategy should compensate for this trend by providing proportionately more direct support for private-sector R&D than in the past, and by maintaining funding of university basic research.

Present Service plans to consolidate R&D activities do not adequately meet the need for a major restructuring of the defense R&D base. A defense R&D base that is smaller and has a new mission will also need a different organizational structure. Current and proposed plans to consolidate the Services' ungainly complex of laboratories and centers were developed before the demise of the Soviet Union. Such plans are therefore unlikely to create the integrated structure that the R&D element of the future DTIB will require. If R&D funding is not shifted to the private sector, the Service laboratories and the Federally Funded Research and Development Centers will have to shoulder much of the responsibility for research and innovation now performed by private companies. If, however, policymakers do shift R&D funding to the private sector, far more Service consolidation will be required. In any event, a smaller DTIB will necessitate greater coordination and consolidation among the Services' R&D efforts and between the Services and the private sector.

The DoD must make greater efforts to exploit civilian technology. Yet without regulatory changes, current performers of military R&D will have no incentive to improve their links to civil R&D. Three areas deserve attention. First, current rules governing independent research and development (IR&D) impose barriers between military and civil R&D activities within companies. Second, current rules allowing the government full rights to corporate technical data developed with government funding discourage specialized subtier firms—a primary source of innovation in defense systems—from developing technologies for both civil and military use. Third, reduced funding will preclude the DoD from maintaining world leadership in all defense-relevant technologies, increasing the need for the United States to benefit from R&D efforts in other countries. Yet current import and export restrictions inhibit interchange between defense and nondefense sectors and prevent the DoD from drawing on technology developed abroad, even by U.S.-based multinational firms.

Current policy places too little emphasis on improving manufacturing technologies. The DoD's effort to develop a Manufacturing Technology Plan warrants support as a way to address the current imbalance between process and product technologies. A remedy will require a new focus on manufacturing technologies by the Service laboratories and the private sector.

Prototyping-Plus

Many people now advocate prototyping in some form to maintain technology innovation during a period when fewer new weapon systems are under development.[10] A DoD acquisition strategy that combines greater use of prototyping and limited production, along with changes in manufacturing and maintenance, might help to preserve critical design and manufacturing capabilities. **As currently practiced, however, defense prototyping is narrowly focused on performance and does not usually incorporate manufacturing and maintenance considerations. As a result, it does not enable defense contractors to move efficiently into production when needed.**

This report assesses a prototyping strategy that combines prototyping with limited production. Termed **prototyping-plus**, it would seek to promote innovation and maintain America's technological edge and its ability to deploy new generations of the most advanced military systems. **Prototyping-plus would involve the continuous development and limited, intermittent production of technology demonstrators and prototypes for operational testing during the periods between full production programs.** By always having some prototype programs under way, the Nation would be in a better position to move the most advanced available systems into production. At the same time, it could maintain a robust weapon design and development capability that could respond flexibly to the uncertainties of the new security environment.

A prototyping-plus strategy is an important part of an overall plan to restructure the DTIB and should break the nearly automatic link between development and production in the current acquisition pipeline. The strategy should be rooted in an understanding of which defense-related design and manufacturing capabilities must be preserved in the absence of ongoing production. Nevertheless, prototyping-plus is not a "research-only" strategy. It includes future force modernization with advanced weapons as needed, after the development and testing of alternative concepts. Some prototyping efforts would aim to develop improved subsystems for upgrading current platforms. Others would focus on developing new platform configurations for potential deployment in the event of a breakthrough in performance (e.g., stealth), the need to replace obsolescent equipment, or the emergence of a large-scale military threat.

Industry has raised a number of objections to pursuing a prototyping strategy. First, some firms contend that while prototyping could preserve design teams, it would involve too few production workers to maintain manufacturing skills. Prototyping-plus, however, calls for the limited production of operational prototypes. Recent trends in manufacturing, such as greater use of concurrent engineering and flexible manufacturing systems, increase the potential to produce limited numbers of prototypes for field testing and to preserve key manufacturing skills without quantity production. The challenge for the future will be to use the construction of a small number of prototypes to identify and correct manufacturing problems associated with quantity production.

A second criticism of a prototyping strategy is that since profits for defense contractors today come from production and not R&D, a prototyping strategy could not keep defense firms in business. This is a valid point, and in the future, prototyping activities will have to be fully funded by the government.[11] Moreover, defense contractors will not rely exclusively on prototype development for their livelihood. Instead, they might derive their income from several concurrent activities, including the low-rate production of new weapon systems; the retrofit, overhaul, and maintenance of deployed military systems; R&D; and prototyping.

[10] The administration, for example, has announced that it will make much more use of prototyping. Congressman Les Aspin, Chairman of the House Armed Services Committee, has proposed a prototyping strategy that he has termed "rollover-plus." OTA proposed a prototyping strategy in *Redesigning Defense*.

[11] Firms would not, however, be precluded from paying for their own prototyping if they believed they had an idea that would "sell," but the incentives to do so might be low.

A third criticism of pursuing a prototyping strategy is that it would have a negative effect on the subcontractors and suppliers at the lower tiers of the DTIB. This objection has some merit because the volume of parts required for prototyping may be too small to keep many subtier firms in business. But new production will not cease entirely in all systems. Subtier firms will build components for continuing (but much reduced) new production, and supply parts for upgrades and retrofits of fielded systems. Many subtiers also have a diversified product line that includes nondefense markets. As a result, the number of subcontractors and suppliers at risk from a prototyping-plus strategy may be small. In some cases, key technologies may have to be acquired by prime contractors or preserved in government facilities. In addition, subtier firms will likely be consolidated into a smaller number of diversified suppliers more closely linked to primes. This restructuring will require changes in acquisition laws and regulations that currently inhibit rather than promote such long-term associations.

Implementing a prototyping-plus strategy would require more integrated DoD management and the reform of defense-procurement laws and regulations. It would also demand a change in mindset by both government and industry from the current focus on producing hardware to a new emphasis on acquiring new technology and know-how as the basis for the Nation's future military potential. A prototyping-plus strategy can help maintain the key design and manufacturing personnel required to develop the next generation of systems. But it should be a part of an overall DTIB strategy that includes continued manufacturing and a viable structure for maintaining and upgrading fielded equipment.

Efficient, Responsive, Mobilizable Production

Redesigning Defense noted that a continued strong production base is essential and suggested three desirable characteristics for future defense production:

1. it should produce weapons and military equipment efficiently in peacetime,
2. it should be responsive to a regional crisis or war perhaps through increased production (''surge''), and
3. it should be capable of greatly expanding production (''mobilization'') in a timely fashion if a large global military threat emerges.

The policies needed to achieve these different characteristics may be in conflict and require tradeoffs. For example, reducing excess manufacturing capacity to promote efficient peacetime production may limit the ability of the base to meet surge requirements in wartime.

The current defense production base has considerable overcapacity when measured against anticipated military requirements. The overall procurement budget may drop by two-thirds (in real terms) from its peak in the mid-1980s. Such shrinkage requires a major restructuring of the production base. *Redesigning Defense* concluded that if this restructuring takes place haphazardly, it could create gaps in critical defense industrial sectors. The government could adopt policies to smooth the transition to a smaller but sounder production base. Alternatives for achieving the desirable characteristics of efficiency, responsiveness, and mobilizability are examined in chapter 4 and briefly summarized below.

Efficiency

Efficient production is defined as manufacturing quality products at an affordable cost. However, for the future DTIB, retaining a manufacturing skill base is also a major stated goal. An efficient defense production base must streamline individual businesses and consolidate industrial sectors. These processes are currently under way. **But unplanned restructuring of the base in response to market forces risks the loss of critical production capabilities, as manufacturers shed important base capabilities such as R&D staffs and training programs.** Policymakers might act to facilitate the consolidation of the production base into a few strong, quality producers rather than retaining many weak firms. To do this, they would need to identify critical producers, modify contracting practices, and change competition rules. (See table 1-3.) In the case of militarily unique sectors, such as nuclear submarines and gun tubes, it may be necessary to support a private or public arsenal to maintain a capability that might otherwise disappear.

The government should ensure that an essential capability continues to exist in the DTIB, but it might be indifferent as to whether a particular

Table 1-3—Policies to Assure Efficient Peacetime Production

Base Structure
- Identify critical producers at the supplier, subcontractor, and prime contractor levels.
- Intervene in the market if necessary to save and strengthen these critical producers through sole-source production, upgrade, spare part, or repair contracts; by removing barriers to mergers and monopoly; or by creating a private or public arsenal.
- Allow consolidation of subtier producers to support innovative, dependable, quality producers.
- Move aggressively toward increased civil-military integration of the base.

Procurement Reforms
- Substitute "best value" or some similar formulation for "lowest bid" as the criterion for contract awards.
- Reexamine the rules on the protection of proprietary data rights.
- Continue the trend towards greater commercial product buys and greater reliance on commercial business practices.
- Reduce costly paperwork, data, and oversight requirements that have been created by law or by the procurement culture.
- Rationalize military specifications to emphasize final quality and performance over production process and use only where necessary.

Acquisition Strategy
- Mandate increased commonality and modularity in systems.
- Increase joint procurement, possibly by moving to "purple suit" procurement, specialized Service procurement, or a civilian acquisition corps.
- Fund the stockpiling or production of items not produced domestically but considered too important to continue sourcing abroad.
- Support low-rate production in critical industries through predictable, multiyear contracts.
- Fund manufacturing skills directly through scholarships, manufacturing technology apprenticeships, the creation of training centers, or indirectly through procurement and incentives.
- Support manufacturing technology developments where appropriate.

International Activities
- Reduce barriers to foreign military sales.
- Support international development and production ventures as a source of technology transfer to the United States.
- Continue to purchase essential supplies and components abroad and determine whether increased reliance on allies is in the national interest.
- Source critical foreign components from multiple countries to avoid cutoff.
- Adopt international specifications and standards where appropriate.

SOURCE: Office of Technology Assessment, 1992.

company continues to produce defense goods and services. The survival of a particular firm or organization need not drive DoD production policy in the long run. Government policies are therefore best targeted toward maintaining capabilities rather than particular companies or government organizations.

Planned low rates of production (unlike the unplanned production stretchouts that have characterized the cold war) can provide U.S. forces with a steady flow of materiel while preserving manufacturing skills, facilities, and equipment that might otherwise atrophy. A DTIB strategy that includes low-rate production will need to establish production rates at an appropriate level: one that preserves the manufacturing complex (primes and subs, private and public facilities) and provides predictable funding so that producers can make major organizational and capital-investment decisions with confidence.

Defense production during the cold war was characterized by plans to equip U.S. forces rapidly with new weapon systems by means of high production rates. In practice, however, budget constraints often lowered actual production rates, resulting in higher unit costs. More realistic future production planning will save money, although it will also reduce surge capacity.

The DoD could supplement low-rate production with prototyping of follow-on systems, spare-parts production, and upgrade and maintenance work. Industrial sectors could be further consolidated so that several related products are built in the same factory (e.g., a variety of armored vehicles or aircraft), a practice common in subtier companies and in some prime contractors. The advent of flexible manufacturing techniques and organization will make this last option more practical over time.

Peacetime production efficiency will be enhanced by lowering barriers between defense and civilian production. These barriers—including special accounting requirements for defense products and stringent military specifications and standards—were created to safeguard public funds and ensure quality. But they also increase defense acquisition costs, place extra burdens on defense companies seeking to diversify into the civil sector, deter leading-edge commercial firms from participating in defense work, and obstruct the flow of technology between the two sectors. A radical solution would be to absorb the defense production base into the civil base, leaving only a few militarily unique products to be built in arsenals. At a minimum, the DoD could continue its efforts to procure more products off-the-shelf and to reduce excessive oversight and specifications through management reforms that shift more responsibility to producers.

Foreign sales of American military hardware can help maintain defense manufacturing and advance U.S. foreign policy objectives. They also carry significant risks. While foreign military sales can maintain U.S. production lines and support allies, sales to unstable or potentially aggressive countries can create new security threats. Collaborative programs with allies help share R&D costs and enable the United States to gain access to foreign technologies. But arms sales often involve "offset" arrangements that give the purchasing country a share of the development and production work, or transfer technologies that can accelerate the proliferation of advanced weapons and may eventually undermine U.S. competitiveness.[12] Purchasing systems, components, and supplies abroad is already a fact of life in an increasingly global economy. But the benefits of lower costs need to be weighed against any risks to security of supply.

Finally, **peacetime production efficiency can be enhanced through manufacturing innovations, including a reliance on common subsystems and parts.** Manufacturers, given the right incentives, can increase efficiency by incorporating new ideas in management, organization, technology application, procedures, and training. Commonality in product subsystems and procurement practices among the Services, if pursued vigorously, would simplify logistics and lower costs.

Responsiveness

A responsive production base is one that is able to react to a crisis or war that is smaller and less demanding than a "total" war demanding national mobilization. **A response to regional threats might be accomplished through some combination of surge production of key items, stockpiles, or reliance on allies.** Each of these three alternatives has strengths and weaknesses. Planning to surge production of materiel when needed avoids the costs of manufacturing and stockpiling. But it entails investment in excess production capacity and thus lowers the efficiency of the peacetime production base. Stockpiled military materiel has the advantage of being available on demand, but it carries manufacturing and storage costs, and it may become obsolete before it is needed. Foreign purchases may cost less but may be susceptible to cutoff or unacceptably long delivery times in crisis or war, and may hinder U.S. development of defense technologies. Moreover, most U.S. allies have small defense industrial bases and are consolidating them. Thus, they may need their own entire output if they are combatants alongside the United States in some future conflict.

Photo credit: DoD

Substantial U.S. military materiel has been prepositioned abroad. Requirements for future conflicts must be met with a combination of production surge capability and stockpiling.

The United States might best focus its surge planning primarily on consumables (e.g., munitions, food, fuel, and spare parts) for intervention in regional conflicts. For the foreseeable future, the U.S. military will probably not require a surge capacity for major weapon platforms and should not fund such a capability.

Mobilization

Responding to a major new military threat on the order of the former Soviet Union would require a mobilization of the Nation's industrial base, as occurred during World War II. Even though the likelihood of a major attack on the United States and its allies is extremely low, the large planned cuts in U.S. active forces would increase the need to

[12] U.S. Congress, Office of Technology Assessment, *Global Arms Trade: Commerce in Advanced Military Technology and Weapons,* OTA-ISC-461 (Washington, DC: U.S. Government Printing Office, Jun 1991).

mobilize if a large-scale military threat emerges in the future. The dedicated defense base would serve as the core of any mobilization effort, supplemented by the domestic civilian production base, war reserve stockpiles, and allied industry.

As a result, **there might be greater emphasis on mobilization preparedness planning and maintaining essential DTIB capabilities.** This task will require a good understanding of the broader national industrial base and realistic estimates of available warning time. Mobilization plans should be reassessed and exercised periodically. Low-rate production, prototyping-plus, and other strategies designed to retain defense manufacturing skills are also central to mobilization preparedness. Policies that foster increased integration of the defense and civilian production bases will aid any future mobilization.

A Robust Maintenance Capability

Depot maintenance, the overhaul of military equipment in specialized facilities as opposed to routine repairs in the field, is critical to the readiness of future U.S. forces. The U.S. defense maintenance base is large ($13 billion in fiscal year 1991). It includes an organic, ''in-Service'' component operated by the different Services that currently performs between 60 to 70 percent of the depot maintenance work. A private-sector component does the remainder of the work and also supplies billions of dollars' worth of spare parts, which are included in the $13 billion. The Nation needs a plan to preserve the maintenance base through the present turbulent period of force reductions, and to restructure it to support smaller numbers of more sophisticated and reliable systems. The most important choices affecting the future maintenance base are:

1. the extent of consolidation,
2. ownership and control of the base (private v. public),
3. emphasis on efficiency v. wartime responsiveness, and
4. the extent to which maintenance can be integrated into the future manufacturing base.

In-Service maintenance facilities were modernized during the 1980s, and there is general agreement that current capacity exceeds realistic future requirements. New DoD initiatives implemented as a result of the 1989 Defense Management Report are streamlining and consolidating the government maintenance base, but there is disagreement about its future structure. The Services generally seek to retain in-Service capabilities, while industry seeks to promote a greater role for the private sector.

Future depot maintenance requirements will differ from those of the past 40 years. Initially, the United States will retire many of its older weapons in response to the waning military threat, reducing the average age of equipment in the field and decreasing the near-term maintenance workload. Over time, however, the lower rate of new weapons production will increase the average age of deployed equipment and make upgrading more important. Future systems will be more sophisticated but also more reliable, changing the nature of the maintenance task and reducing maintenance requirements.

The ongoing consolidation of in-Service depot maintenance (including single sites for each technology and significant reductions in workforce) is a major achievement by past standards, but is still insufficient to meet the needs of the new security environment. Performance of maintenance tasks across Service lines remains limited. Moreover, despite reductions in manpower and consolidation of workload, the maintenance base contains almost the same number of major facilities as existed to support a defense establishment prepared for war with the former Soviet Union. While the drive for peacetime efficiency must be tempered by the need for responsiveness in a major future crisis, the foreseeable demand for wartime maintenance support has greatly diminished with the end of the cold war.

Private industry has the capability to do more depot maintenance work and is eager to assume this role. Proponents of transferring more maintenance work to private firms note that they already possess an inherent maintenance capability by virtue of having manufactured the equipment. Further, manufacturers typically provide depot maintenance support until a system has been deployed in sufficient quantity to permit standardization of maintenance procedures by the military user. Since the manufacturing firm has already developed test equipment and trained personnel, there is an additional cost (depending on the particular system) in developing a separate Service maintenance capability. Proponents also argue that private firms are more efficient than government depots.

Photo credit: Lockheed Aeronautical Systems

A private company maintains ejection seats and other components for the T-33 jet trainer built in the 1950s. As production funding declines, private industry is increasingly interested in maintenance work.

Those who favor retaining Service maintenance capabilities contend that shifting maintenance to private firms would reduce Service flexibility, increase the risk of inadequate responsiveness in wartime, and leave the DoD vulnerable to cost escalation. Yet there is no clear evidence that private firms cannot be responsive or that private-sector costs cannot be controlled. Nor is there clear evidence that private firms are inherently more efficient than Service depots. Accordingly, there is a need for more study of the proper private-public mix in the future maintenance base.

During the downturn in production, maintenance might play an important role in supporting manufacturing capabilities in industrial sectors where there is an overlap in processes, equipment, and skills required for manufacturing and maintenance. In many cases, maintenance and upgrades could be carried out in the same factories as new manufacturing. Some industries, however, have little overlap between manufacturing and maintenance. In such cases, combining maintenance and manufacturing would require restructuring the production process to take advantage of synergies between manufacturing and maintenance on the factory floor.

Congressional actions have made the rationalization of the depot maintenance base more difficult. Legislation limiting competition, directing work to particular facilities, and mandating job protection have all constrained the DoD's ability to operate the maintenance base efficiently. Properly sizing the future maintenance base will require a broader view of overall DTIB requirements and decisions designed to support the integrated base rather than its individual parts.

Good, Integrated Management

Good, integrated management is fundamental to the successful operation of the future DTIB. Such management must anticipate future needs and take action to ensure that the base can meet them at an affordable cost. Good management does not imply any particular amount of direct government intervention in the DTIB, but it does allow for intervention if needed to ensure the survival of a critical technology or industrial capability.

Future DTIB management could be integrated with respect to the three functional elements of the base (R&D, production, and maintenance), the three Services, the Executive Branch and Congress, and government and industry. Peacetime procurement could also be integrated with crisis and war planning. Integration among the R&D, production, and maintenance elements of the base would ensure that managers understand how these three elements interact and make decisions optimized for the entire base rather than an individual element or subelement. Integration of DTIB planning within the DoD and the Services can eliminate redundancies in Service capabilities (e.g., laboratories and depots) and Service-specific contractors. Several DoD initiatives are addressing these issues, but there is still much resistance to closing and consolidating facilities. The difficulty of consolidating the current base would be eased through better coordination among the Executive Branch, Congress, and private industry, since DTIB management is ultimately a national, rather than a DoD, responsibility.

Photo credit: Defense Systems Management College

The Defense Systems Management College trains government acquisition personnel responsible for billions of dollars of purchases a year.

The immediate management challenge is to plan rationally for the major shrinkage and restructuring of the DTIB. Three strategies for approaching the transition to a smaller base are discussed in chapter 6. The Nation could employ:

1. a **free-market strategy** that relies on market forces to decide which defense contractors and government facilities will survive,
2. an **activist strategy** in which the U.S. government attempts to manage the change by identifying critical firms and facilities and ensuring their survival, or
3. a **mixed strategy** that allows market mechanisms to operate when possible but uses government intervention to preserve critical defense industrial capabilities that might otherwise be lost.

Applying exclusively a free-market approach to DTIB management is likely to result in a weakened and inefficient base as firms shed capabilities to remain competitive. Yet the DTIB is too complex to allow detailed centralized control. Thus, an optimal approach might combine centralized planning with decentralized execution. For 40 years, the DTIB has been an increasingly regulated market with a single government buyer, so that free-market forces are unlikely to operate efficiently. Because the DTIB is part of the larger national industrial base, however, it can potentially take advantage of market forces within the larger base. Modifying acquisition laws to open up the DTIB to a larger number of companies would enhance the effects of market forces. But if policymakers choose to retain the current acquisition system, more government intervention may be needed to ensure that crucial elements of the base are preserved.

Laws, regulations, and bureaucratic behavior inhibit DTIB managers from making greater use of the civilian technology and industrial base. Future base managers will need to be more creative in using the entire range of potential technology and industrial resources, including civil and foreign firms, rather than concentrating on dedicated defense producers. Key to successful management of the future base will be the purchase of commercial products, the use of civilian production facilities, and the adoption of commercial operating procedures. Achieving such civil and military integration will require less day-to-day DoD and Service control over technology and industrial assets and an increased ability to make use of the wide array of goods and services existing in the civilian sector.

THE FUTURE INTEGRATED BASE

The previous sections described desirable characteristics of the future DTIB and suggested some alternative strategies for achieving each. **These characteristics cannot be viewed in isolation, but must work together for the future base to be healthy and meet the Nation's security needs.**

OTA's analysis suggests that, given the likely reductions in defense budgets, minor changes in the structure and operation of the DTIB will not suffice to provide the Nation an effective future military capability. One of the principal findings of this assessment is that the base can be restructured to exploit the inherent synergies that can result from

Figure 1-2—The Future Integrated DTIB

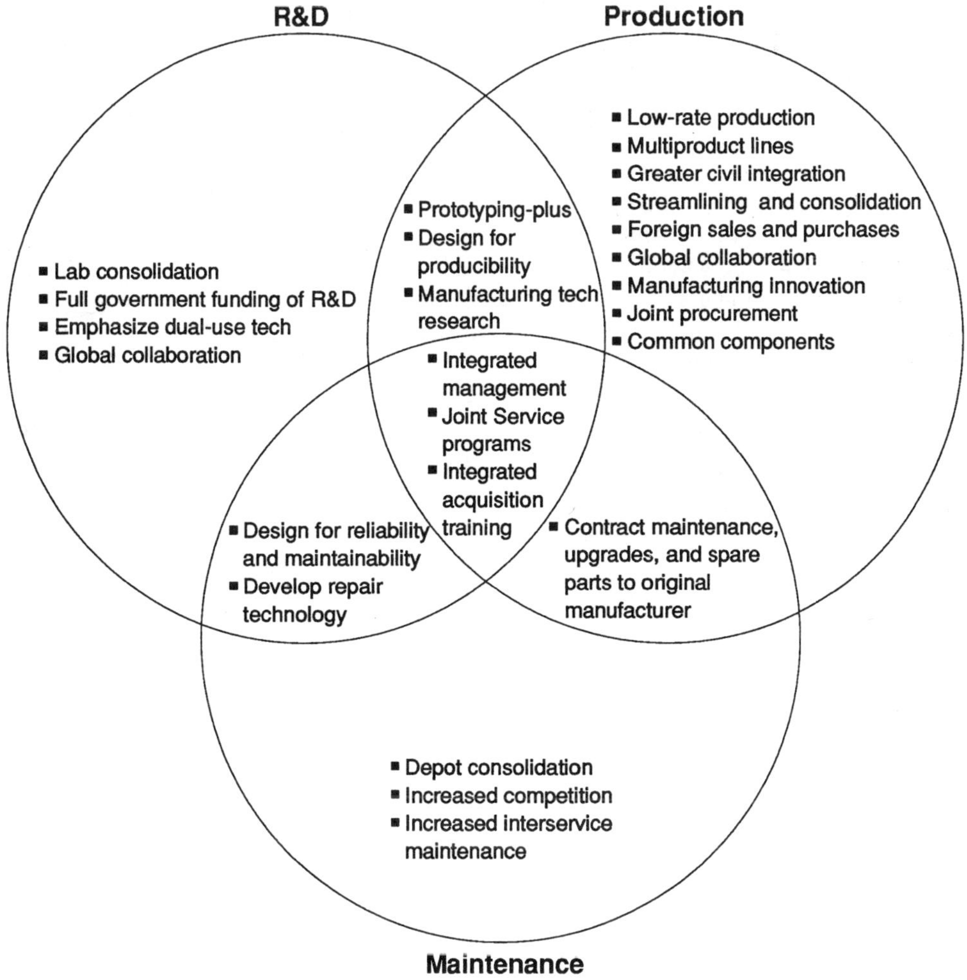

SOURCE: Office of Technology Assessment, 1992

closer integration of the R&D, production, and maintenance elements.

The interrelationships among the three functional elements of the future DTIB—R&D, production, and maintenance—are portrayed in figure 1-2. Although these elements are largely managed separately in the current base, the figure suggests that the three elements could be structured and managed in an integrated manner to yield greater efficiencies. For example, R&D could be directed not just to creating new products but to making the manufacture and maintenance of those products as simple as possible. Similarly, carrying out production and maintenance activities in the same factories would facilitate low-rate production.

Integration could also be carried across Service lines and between defense and civilian industry. The DoD could enforce joint Service use of common equipment (for example, air-defense systems) and ultimately eliminate all barriers between the civilian and military technology and industrial bases. It could move to a centralized acquisition corps separate from the Services, perhaps along the lines of the French model. A fully integrated industry might handle R&D, production, and maintenance, with very little of the base remaining under government ownership. Such a radical restructuring of the DTIB would require a substantial change in the attitudes of both government and industry.

Chapters 2 to 6 consider ways to exploit the synergies among activities, including prototyping-plus, low-rate production of selected military equipment, restructuring of assembly lines to be more diversified and flexible, increased civil-military integration, and more competition in all three elements of the base. Yet if each of these alternatives were pursued in isolation, it would have only a modest impact and might even be detrimental to the base as a whole. For example, critics have argued that prototyping alone cannot maintain an effective manufacturing capability and would not be profitable enough to keep defense firms in business; that low-rate production of new systems would fail to provide an adequate economic return, result in high unit costs, and lead to failures among subtier producers; and that shifting more maintenance work to the private sector would reduce the responsiveness of the base to military requirements.

None of these criticisms lacks validity. Although the options suggested in this report all entail risks, so do current DTIB policies if they are simply extended into the new era of reduced budgets. The critical question is not whether the DTIB will shrink, but how best to restructure the base to assure the Nation's future security.

The following section describes the general characteristics and management activities of a future integrated DTIB. Boxes 1-B to 1-D contain a hypothetical scenario set in the year 2010 illustrating synergies among the elements of the DTIB and the implications of alternative policy choices.

National-Level Decisions

National decisions on overall defense funding and funding priorities are based on assessments of the military threat, as well as economic and political conditions. Once the total resources to be directed to defense have been established, DTIB managers at the national level allocate them to the various national security goals and elements of the base. These funding decisions involve the strategic choices outlined earlier in figure 1-1 and require supporting policies if they are to succeed. Examples of policies associated with each of the choices are shown in figure 1-3.

In the post-cold war era of diminished immediate military threats and reduced budgets, a healthy future DTIB requires shifting funds from current production to R&D. The policies outlined in figure 1-3 indicates that a healthy base also requires a commitment to purchase *knowledge* rather than *hardware*. As noted earlier in the discussion of figure 1-1, a shift of resources to future-oriented investments will not be universal, since some systems will need to be produced to replace obsolete equipment or to respond to an emerging threat.

A decision to emphasize dual-use technologies or civil-military integration would require the DoD and the Services to increase reliance on commercial firms, provide incentives for using non-developmental items, and stress performance criteria over rigid military specifications. These policies would require greater initiative on the part of government contracting officers than is currently allowed, and therefore better trained government acquisition personnel.

A decision to retain strong private-sector involvement in the DTIB, instead of letting most activities in the base devolve to the Services and the DoD, would require rule changes that enable industry to obtain profits commensurate with risks. Although the overall strategy stresses the private sector, the DoD may still have to maintain some critical capabilities in GOCOs or GOGOs.

The choice of single-source or competitive procurement will be driven by demand and market structure. Policies to support dual-use technologies and civil-military integration would increase the participation of commercial firms and thereby strengthen competition. While militarily unique items might also be acquired competitively, sole-source procurement may be preferable to artificial competition in those areas where civil-military integration is not feasible.

Finally, the United States has the choice of drawing on international sources of technology to enhance its own military capabilities. The ability to do so would be facilitated by negotiated international agreements that promote reciprocity in defense trade. Nevertheless, domestic sources for some critical defense-related items should be preserved. The policy questions are, "which items?" and "how to preserve them?" Improved databases are essential to answering these questions.

Despite significant reductions, the actual levels of future defense funding are very uncertain. All estimates indicate that as long as the United States seeks to remain a major world power—let alone the preeminent one—defense spending will remain at a fairly high level. Table 1-4 shows some possible DTIB funding allocations for the first decade of the

Figure 1-3—Policies Supporting Future DTIB Strategic Choices

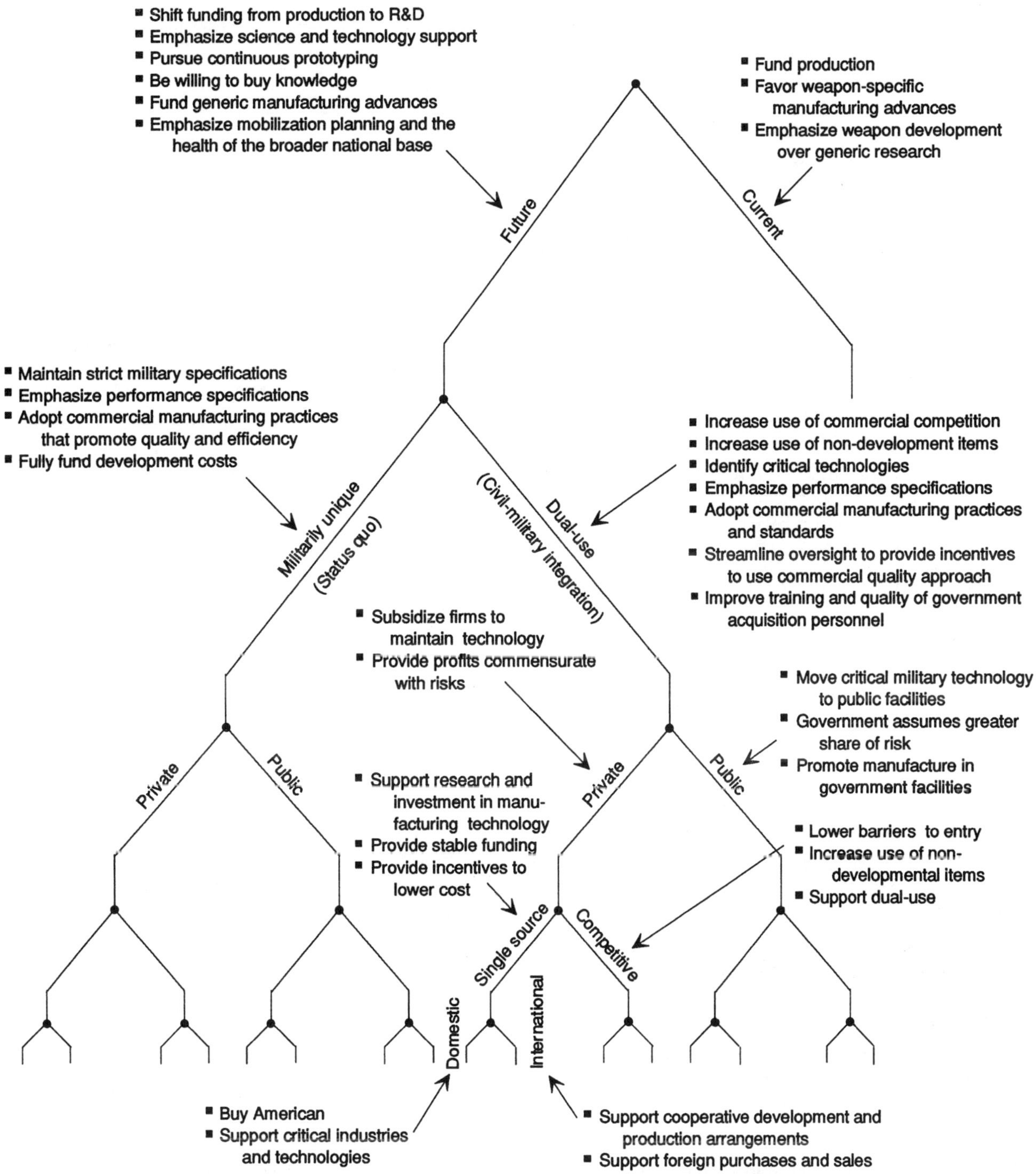

SOURCE: Office of Technology Assessment, 1992

Table 1-4—Hypothetical Annual DTIB Funding Alternatives for 2001-2010[a]
(billions of fiscal year 1992 dollars)

Total DoD budget	Total DTIB funding	R&D	Prototyping[b]	Production	Maintenance
220	88	24	9	47	8
180	72	20	10	35	7
150	61	17	11	27	6

[a] Estimates are based on cold war allocations with adjustments for changes in the nature of the military threat.
[b] As overall production declines, prototyping funding is increased to maintain technological innovation, key design skills and some manufacturing techniques.

SOURCE: Office of Technology Assessment, 1992.

21st century, assuming a reduced global military threat and a shift from full-scale production toward greater use of prototyping. Assuming reasonable U.S. economic growth rates, these budgets might represent expenditures of 2 to 3 percent of the GNP.

The table suggests that even with major budget cuts, DTIB spending might be adequate to support a significant defense R&D, manufacturing, and maintenance effort. But DTIB restructuring would be needed and within these funding constraints, force modernization decisions would be made at the DoD, rather than the individual Service level.

The hypothetical scenario in boxes 1-B to 1-D reflects a policy emphasis on future over current capability, dual-use over militarily unique technology, private over public-sector facilities, competitive over sole-source procurement, and international over domestic sourcing, where international is defined as choosing the best technology regardless of source. (See box 1-B.)

Industrial Sector Strategies

A national strategy to restructure the DTIB would be implemented differently in the various defense industrial sectors because the sectors differ in:

1. their degree of integration with the Nation's industrial base (e.g., electronics is more integrated than ammunition);
2. their economic health (e.g., the aircraft industry is healthier than shipbuilding); and
3. the amount of military goods and services the military buys from the sector in peacetime.

To maintain at least one source of design, development, production, and maintenance for each system and component, the government may have to compromise on weapon performance and make significant changes in acquisition laws and regulations. Indeed, in some sectors the demand may be so limited that a single-source arsenal (public or private) may be required to preserve the technology. (See box 1-C.)

DTIB managers will need to look for synergies to reduce overall costs and improve efficiencies. For example, research on common technologies may be consolidated among the Services. It will also be important to identify bottlenecks and gaps in the DTIB so that remedial actions can be taken. To achieve a small but flexible defense base, managers will need a better overview of industrial capabilities and potential than has existed in the recent past.

Organizational Implications

Companies that decide to stay in the defense business may have to make significant internal changes. These include:

1. concentrating on a defense market niche or, alternatively, becoming a full-service defense firm with high-quality design, production, and maintenance capabilities;
2. streamlining;
3. expanding horizontally or vertically into new military product lines or, alternatively, concentrating on current lines; and
4. better integrating military work with civilian work and/or expanding into the civil sector.

Any move to low-rate production suggests that firms will probably need to manufacture more than one product and engage in some maintenance activities. To give firms an incentive to move in this direction, the DoD will need to change its contracting criteria and acquisition rules, and might also fund innovations in manufacturing technology such as multiproduct assembly lines. As the DTIB moves toward greater civil-military integration, the DoD may also have to modify weapon design in order to make better use of the civilian base and take advantage of commonalities in systems. (See box 1-D.)

Box 1-B—National Strategy in 2010

Beginning in the 1990s, the United States undertook a major shift of resources from defense to other national priorities as a result of the reduced global military threat, fiscal problems, and the need to strengthen the economic foundation of national security. U.S. military forces have been involved in a few limited military operations over the past decade, but no major new military threat has emerged. Defense spending for 2010 is $180 billion in fiscal year 1992 dollars. DTIB funding as a percentage of overall defense spending has increased in relative terms, but has fallen in absolute terms.

Allocation of DTIB Funding in 1991 and 2010

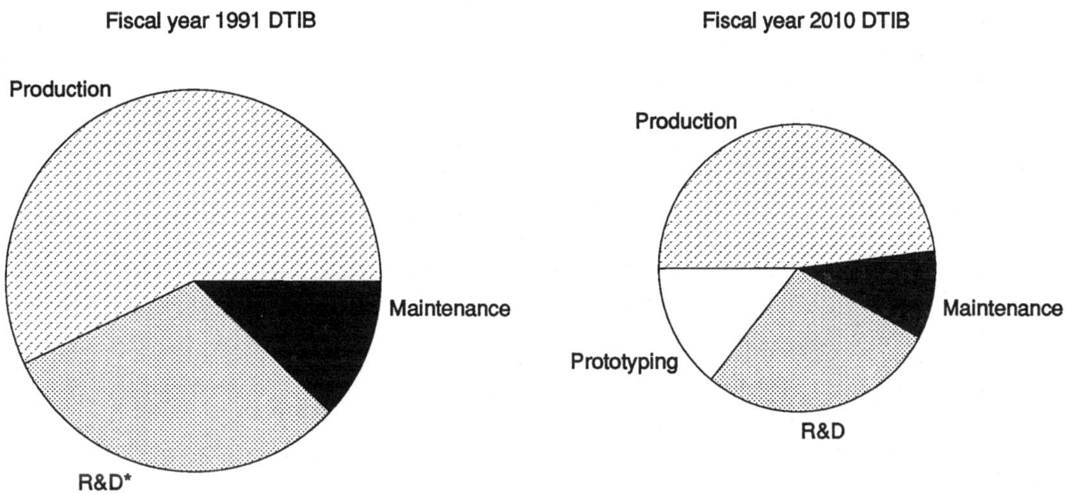

*Prototyping is included in the fiscal year 1991 R&D funding.

SOURCE: Office of Technology Assessment.

The United States remains not only the strongest global military power, but also the one having the greatest military potential. With limited funding, the U.S. DTIB strategy emphasizes three principal thrusts. First, DTIB spending generally emphasizes maintaining military potential rather than current capability. A prototyping strategy is being pursued across the weapons spectrum. This strategy includes a shift in emphasis from preproduction prototypes to the use of computer simulations, technology demonstrators, and low-cost prototypes to test new concepts. The $10 billion prototyping budget maintains several design teams in critical defense areas such as high-performance aircraft, ground combat vehicles, and new munitions, as well as for the myriad of subsystems and components that go into these systems. R&D is receiving a relatively large share of DTIB funding compared to 1991, and more funding has been dedicated to manufacturing and maintenance technologies.

Second, even with the emphasis on military potential, production remains the largest single component of DTIB spending at $35 billion—a drop of about 40 percent since 1991. Current production includes end-items (e.g., ships, aircraft, armored vehicles, and munitions), their embedded components (including upgrades), spare parts, and prototypes. The DoD is following a systematic approach to force modernization and continuing to replace weapon systems as they become obsolete. But, because of the limited production funding, decisions on major new weapon programs require more joint-Service analysis and cooperation to achieve national, rather than individual Service, objectives. A significant percentage of procurement funds go to upgrading older fielded weapon systems to improve their overall capabilities.

Third, the defense acquisition process and the DTIB have been restructured to make extensive use of civilian industry. For example, defense R&D administrators leverage their $20 billion budget by focusing in-house efforts on militarily unique technologies and by assimilating or adapting new civilian scientific and technical developments to meet defense needs. Weapon and component designs increasingly incorporate commercial products and processes, and military and civilian products are often manufactured side-by-side. Private firms are heavily involved in providing depot level maintenance and upgrades for deployed forces.

Box 1-C—Armored Vehicles and Helicopters in 2010

Armored vehicles and helicopters remain important components of the United States' combat capability, but they are evolving over time. There are about 13,000 armored vehicles and 2,000 helicopters in the active U.S. inventory. Most are older, but some new systems have been introduced during the last 15 years.

Strategies to maintain the two sectors have both commonalities and differences. Attempting to maintain a reasonably modern force (vehicles and aircraft not more than 30 years old) at current force levels requires an average production of several hundred new vehicles and more than one hundred new aircraft each year. Actual production is limited by available funding. Armored vehicle production is receiving a little less than 3.5% of the fiscal year 2010 procurement budget (about $1.23 billion in fiscal year 1992 dollars), and helicopter production, with higher priority, is receiving about 5.5% ($1.93 billion) of the procurement budget. These funds would be insufficient to maintain the current force at the unit cost levels existing in the early 1990s.

The national DTIB strategy stressing civil-military integration of the base (i.e., reducing unnecessary military specifications and purchasing components and using processes from the civil sector) has helped reduce unit costs, but not enough to maintain the desired force structure at current funding levels. The DoD is examining additional ways to lower unit costs, the possibility of increasing the relative share of production funding for these two sectors, or further reducing force levels.

System upgrading is a major component of each sector strategy. Upgrades include new electronic components with improved reliability, improved night-vision systems, fire-control electronics, and antitank missiles in both helicopters and armored vehicles. The consolidation of R&D in some critical defense technologies (e.g., optoelectronics and advanced materials) has enabled the DoD to maintain a world-class effort in these areas even with reduced R&D funding.

Prototyping is a particularly important part of the armored vehicle sector strategy. Funding is divided among computer simulations, technology demonstrators to explore new technical concepts, and the development and testing of operational prototypes. The Army continues to develop prototypes of lighter weight armored vehicles. Component prototyping efforts have been the backbone of all the weapon system upgrades that have occurred over the last decade. DoD policy emphasizes using private firms with production facilities for the design and prototyping, but some prototyping is occurring in specialized "design firms." Contracts for operational prototypes require production of these operational prototypes on standard flexible manufacturing tools, thereby favoring organizations with manufacturing capabilities or engineering firms linked to manufacturing firms. Research and development on some militarily unique technologies is being conducted in government laboratories and arsenals.

The DoD has a stated policy of maintaining more than one source for design, development, production, and maintenance for each system and component wherever feasible. To date it has been able to maintain this policy in both these sectors, but the DoD has sometimes been forced to compromise on initial performance criteria to increase purchases from companies developing similar equipment commercially. New armored vehicle and helicopter designs now have many more common components with other vehicles and helicopters than in the past.

New design concepts have been encouraged by DoD investments in flexible manufacturing through government manufacturing technology programs and investment incentives, and through changes in procurement rules that have cut back direct government oversight. The government has made it easier for firms to use defense production lines and machines for non-DoD work; as a result, the DoD can support more than one prime contractor and several sources for many components (though some components are single sourced). Changes in acquisition procedures, and the replacement of many military specifications by international standards, have increased the number of potential producers of military equipment. Foreign sales remain an important but relatively small part of overall production.

Although both sectors have fewer defense prime contractors in 2010 than in 1991, there are more subtier firms with the potential to provide components to the defense sector. Further, the surviving defense firms remain strong and capable of developing and producing future systems.

Box 1-D—Operations at Selected Defense Firms in 2010

Both private defense contractors and government facilities have been restructured to support smaller U.S. military forces. Surviving prime contractors have either consolidated their manufacturing of similar products into single, privately-owned facilities (sized to meet expected peacetime production needs and lacking excess surge capacity) or have become managers of government-owned facilities—also sized for peacetime rather than wartime production. Subtier firms usually run integrated civilian and military production lines. Consolidated armored vehicle and helicopter production facilities manufacture as many as 3 to 5 different types of armored vehicles and 2 to 3 types of helicopters. Income from low-rate production of several systems is supplemented with spare parts production, repair and overhaul work, prototyping and the continuous upgrading of older equipment. Multiyear contracts provide greater predictability of cash flow and enable the firms to make long-term investments and establish links with subcontractors and suppliers.

Developments in manufacturing technology (such as flexible or ''agile'' manufacturing) have aided the restructuring process. Firms are employing multidisciplinary engineering teams to develop prototypes that are built in their regular production facilities. Prime contractors have relatively more design and engineering capability than in the 1980s. Firms are less concerned with the yearly production of any single system than with maintaining adequate levels of production of several different systems over several years.

Summary

A defense establishment funded at $180 billion (fiscal year 1992 dollars) or even at $150 billion, as shown in table 1-4, will require first-rate technology and industrial support. The portion of the defense budget allocated to the DTIB may fall to the $55 to $70 billion range (fiscal year 1992 dollars). Though considerably smaller than today's DTIB spending, this level of industrial activity would remain a significant national investment. This investment can only be used effectively, however, if the DTIB is restructured successfully through the collaborative efforts of the White House, Congress, the DoD, and industry.

POLICY ISSUES FOR CONGRESS

The DTIB described in both *Redesigning Defense* and this report is complex. Although it is best understood by breaking it down into its component parts, the base can only be managed effectively if it is viewed as an integrated whole and if decision-makers take actions optimized for the entire base.

Policy issues concerning DTIB restructuring in which Congress has particular interest fall into three areas. The first involves **funding**, both total DTIB funding and the funding mix within the Federal budget. The second involves **organization**, including restructuring the institutions in the current DTIB, integrating them with the civilian base, and improving the ability of private firms to meet future defense needs. The third involves **management** of the transition and improving the DoD's coordination of the critical elements. A key issue is how qualified DTIB managers can be recruited, trained, and retained.

Funding

Congress has the constitutional responsibility to provide for the Nation's defense. The decline in the Soviet military threat permits major reductions in defense spending. The administration estimates that defense spending will fall to $237.5 billion (fiscal year 1992 dollars) by 1997. This level corresponds to about 3.4 percent of the gross national product, the lowest percentage in the past 50 years. Many Members of Congress advocate even deeper cuts. Whatever the level of overall defense spending, Congress will also have to make a judgment on the appropriate level of DTIB funding within that budget.

Funding for the DTIB should reflect the fact that it is a critical component of U.S. national security. The DTIB is vital both to the ability of U.S. forces to handle regional military threats and as a hedge against a reconstituted global threat. During the cold war, the combined R&D and procurement budgets averaged about 36 percent of the DoD budget. This share fell after the Korean and Vietnam conflicts but rose by almost 10 percentage points during the military buildup of the early 1980s. The current DoD budget request envisions DTIB funding in fiscal year 1993 slightly below the cold-war average.

In the post-cold war era, with a requirement for fewer active forces and the potential for greater leverage of military forces through technology, Congress might consider giving the DTIB a relatively larger share of the defense budget than has been the case in the past. Since such funding would compete with force readiness, it would probably not be attractive to military leaders. Nevertheless, maintaining relatively high funding would provide modern weapons and support to smaller U.S. forces, assure them a technological advantage in the field, and hedge against future threats. A smaller, better-armed force is preferable to a larger, less-well-armed force.

Congress will also have to consider whether to continue funding defense R&D at a high level. An alternative would be to limit defense R&D to a relatively small number of militarily unique technologies, relying on civilian R&D, perhaps government funded, to generate dual-use technologies with defense applications. Congress should also consider whether to fund manufacturing technology through the DoD, with the possibility that side benefits will filter into the broader industrial base, or to fund such efforts in the civil base and let defense production draw on that larger commercial technology pool. If the Nation adopts a strategy for strengthening civil technology, as several recent studies have proposed, defense could also benefit.[13]

Within the overall DTIB funding level, the appropriate allocation among the R&D, production, and maintenance elements of the base is critically important. A prototyping strategy, whether implemented as proposed by the DoD or along the lines of chapter 3 of this report, would take funds away from production. Prototyping should anticipate no automatic connection between the development of a prototype and a decision to go into quantity production. Congressional debate is likely to revolve around the wisdom of spending relatively large sums of money on prototyping programs that may yield little operationally useful hardware for extended periods. Further, since companies are unlikely to invest their own funds in developing prototypes that have no immediate prospect of entering production, the government's share of the bill may appear relatively large compared with the past. Prototyping under these conditions often involves buying knowledge rather than hardware.

Despite the shift toward R&D, it will be essential to maintain production capabilities in the future. Cancellations or stretch-outs of ongoing and planned procurement programs will shrink the production base and may leave some manufacturing facilities with no production contracts for several years. As a result, **Congress will have to consider funding options that maintain key manufacturing skills and facilities during a period when few new systems are produced.** Greater civil-military integration, if pursued, will require legislative changes.

Funding options for future defense production include:

1. low-rate production spread over multiyear procurements,
2. intermittent production at higher production rates followed by laying away production lines,
3. funding international collaborative production programs,
4. increased foreign sales,
5. innovations in manufacturing technology,
6. designing for more commonality in systems, and
7. the use of single sources to gain economies of scale.

These options can leverage limited funding, but all have drawbacks. Low-rate production may increase unit costs (although the increases might be limited if facilities are kept small and the product is designed for low-rate production); intermittent production can result in the dispersal of workers during periods when there is no production; collaborative programs can involve the partial loss of technology to foreign competitors; and foreign arms sales may provide weapons to new military threats.

Government funding for production may use some mix of these approaches, combined with funding for prototyping and maintenance. Changes in manufacturing funding that encourage firms to produce multiple products may help make low-rate production a more effective tool for maintaining the production base. The DoD can encourage this shift

[13] A Carnegie Commission Study, *Technology and Economic Performance*, September 1991, recommended that the Defense Advanced Research Projects Agency (DARPA) be refocused into a National Advanced Research Project Agency (NARPA) to provide stronger links between military needs and commercial industry. A subsequent report by the Hudson Institute recommended the establishment of a National Technology Agency.

by funding manufacturing advances to assist defense firms that seek to produce multiple products.

Funding for maintenance will also decline, but probably not as much as for production. Maintenance funds are currently spent mainly in the public sector. **Congress will want to consider the most effective future mix of public and private maintenance depots.** Increased competition is expected to lower maintenance costs, but competition is currently concentrated on work traditionally available to the private sector. Meanwhile, the government sector retains a large core of work that is not open to competition. Both the role of competition and the future size of the in-Service maintenance core should be examined in detail.

Congress will want to consider ways to retain people who are critical to the strength of the DTIB. Policy options include predictable defense funding, which can provide the basis for longer term personnel planning; support for technical education and apprenticeships that benefit both the DTIB and the broader national industrial base; support for engineering education in relevant technologies; a prototyping strategy that maintains design teams as well as innovation; and some continued defense production. The recent Defense Acquisition Workforce Improvement Act was an important step toward improving the contracting and program-management capabilities of the defense acquisition process. Corresponding steps are required to ensure the technical competence and overall management of the DTIB.

Organizational Changes

Congress will face a number of organizational changes in the DTIB. Some of these changes are internal to the traditional base, while others are external to it. The external structural changes appear to be the more important. *Redesigning Defense* concluded that the DoD faces the choice of greater integration with the civilian industrial base or maintaining a defense-unique base that will most likely devolve to a set of sole-source providers (''arsenals'') in the public and private sectors. Several studies have found that increasing the integration between military and civilian technology and production will lower overall defense costs, promote technology transfer, increase available industrial capacity, and strengthen the economic dimensions of national security.[14] OTA's discussions with industry and government personnel support these conclusions. The expected deep reductions in defense spending make civil-military integration all the more important.

Moving toward greater civil-military integration will, however, require Congress to make major policy changes in a number of areas. **First, it will be necessary to amend the Federal procurement laws that have tended to isolate the DTIB from the broader base.** *Redesigning Defense* outlined some of these laws, and chapter 4 of this report lists areas of additional concern. The DoD Advisory Panel on Streamlining and Codifying Acquisition Laws is expected to make significant recommendations in January 1993 for simplifying the acquisition laws, and Congress will want to consider these recommendations carefully.

Second, the DoD's ability to increase purchases of commercial and nondevelopmental products depends on reform of acquisition laws and modification of the military specifications that control most defense manufacturing. One approach is to accept commercial and international standards in place of military specifications. The DoD could make a concerted effort to implement standards that, in addition to serving defense needs, help make U.S. firms more competitive internationally. The inability of the United States to accept the metric system is one indication of the difficulty of implementing new standards.

Third, the shift toward greater civil-military integration may require substantial changes in defense R&D. As noted above, Congress might restrict funding for militarily unique R&D and shift more funds to research on dual-use technologies of both military and commercial interest, perhaps by creating a new agency for promoting technological innovation in the civil sector.

Fourth, in the absence of a shift toward greater civil-military integration, Congress will have to consider ways to assure the benefits of competition in a smaller DTIB that has fewer sources of supply. This might entail allowing more competition with allies.

---

[14] These studies include two Defense Science Board Reports on the use of commercial items for defense, a Defense Science Board study on the defense industrial base, and a study by the Center for Strategic and International Studies on civil-military integration.

326-447 - 92 - 2 : QL 3

The second broad organizational issue facing Congress concerns the internal restructuring of the DTIB. Congress will want to examine the consolidation now going on within and among the Services, in industry, and between the private and public sectors. The public-private split that is appropriate will be influenced by the degree of civil-military integration. Congress might promote more rational consolidation by supporting multi-Service procurement and increased inter-Service maintenance for equipment and supplies, and by providing firms an incentive to maintain R&D as well as production capabilities. Changes will also be needed in multiyear procurement rules and the Competition in Contracting Act to promote long-term association between prime contractors, subcontractors, and suppliers.

Management Options

The immediate task facing Congress is to ease the transition to a much smaller post-cold war DTIB. Although the administration has generally advocated letting the free market shape the DTIB, it has also expressed concern over the need to preserve some components of the base. In recent testimony, for example, Secretary of Defense Cheney specifically cited shipbuilding as a problem area.[15] Congress will need to consider the degree of intervention that is appropriate to downsize the base in a rational manner. Chapter 6 observes that increased civil-military integration of the base, accompanied by changes in acquisition practices, could increase free-market competition.

The best approach to restructuring appears to be a mixed strategy that fosters true competition wherever possible (enhanced by greater use of the civilian base) and limits government intervention to those cases in which there is no alternative. But such a strategy would require good information on current DTIB capabilities and future requirements. To this end, Congress might establish and fund a joint legislative-executive commission that would report to the President within 6 months concerning the current capabilities of the base and future requirements and provide some overall guidelines for the downsizing of the DTIB. There is also a need for a more systematic approach to DTIB data collection over the long term. As the future defense base becomes more integrated with the broader civil base, the DoD might not be the best agency to maintain this information, and Congress might consider alternatives such as the Department of Commerce.

Finally, Congress will want to consider how best to balance efficiency and accountability in the future DTIB. Although accountability in the use of taxpayers' money is essential, the issue is how to achieve it most efficiently. Increased civil-military integration has the potential to impose market discipline on more producers, but not necessarily on manufacturers of militarily unique products where accountability will probably still require administrative oversight. Although the DoD has had numerous programs to increase contractor responsibility, the programs have largely failed because of inadequate support or lack of incentives. Congress might consider ways to improve the effectiveness of such efforts.

Ultimately, good management will depend on recruiting and retaining skilled and experienced DTIB managers. Recent steps to improve education and pay are helpful, but Congress should monitor these activities to ensure an improvement in the quality of management personnel.

SUMMARY

This report analyzes the desirable characteristics of the future DTIB described in *Redesigning Defense* and considers alternative policies for achieving them. Restructuring the DTIB will require managing the base as a whole rather than allowing managers in the individual elements (R&D, production, and maintenance) to pursue policies optimized for their separate benefits. Achieving a strong and healthy future base will require an overall strategy that properly considers the trade-offs between investment in current capability versus military potential. Currently these trade-offs are being debated in terms of continuing to invest in current products versus moving funds to research and development.

If DTIB planners look beyond the next decade they will see that even in a relatively peaceful world the Nation will need an effective defense base. The DoD and Congress can plan the transition to a smaller but robust base by emphasizing military potential over current capability. Such a strategy must be applied with care and include limited production of new products to permit force modernization and avoid the erosion of manufacturing

15 "Cheney Opens Door for Steps to Preserve Industrial Base," *Aerospace Daily*, Feb. 3, 1992, p. 174.

expertise. Overall, the changes in the military threat facing the Nation provide many opportunities and challenges for Congress. The ability to transfer funds to nondefense priorities is a great opportunity. Deciding how best to spend the remaining defense dollars is a great challenge.

Chapter 2

Research and Development

Contents

Figures

Chapter 2
Research and Development

INTRODUCTION

The Nation and the Department of Defense (DoD) have a long-standing policy of support for a vigorous research and development (R&D) program to meet immediate and long-term defense needs. Two of the desirable future DTIB characteristics described in *Redesigning Defense* are an advanced R&D capability and greater exploitation of civilian R&D.[1]

The national R&D base has had many successes in providing the military Services the technology they need, although questions persist about the efficiency, direction, structure, and size of defense R&D. These questions include concerns about whether the Nation is pursuing the correct technological paths (e.g., process or product technology) and whether the defense R&D effort is properly organized (e.g., among Service R&D efforts).

There is a broad national consensus today, reinforced by the Persian Gulf War experience, that the United States should preserve qualitative superiority in weapons performance. Unresolved issues that Congress should consider are: how great this qualitative superiority should be, how should it be maintained, and against what threats should it be measured? The most cautious alternative is to continue to compare U.S. weapons across the board to the best capability of any potential adversary. Given the waning military threat from the former Soviet republics and the global arms trade, this may mean U.S. weapons would be compared to the best systems available on the international market.[2]

Even with the world's best weapon performance as a benchmark, the magnitude of the desired performance edge remains a matter of debate. Some analysts argue that the United States should maintain a military advantage sufficient only to defend core national interests and that the weapons already in the pipeline have capabilities in excess of foreseeable needs; therefore defense R&D could be reduced or shifted largely to civil objectives. Other analysts counter that there will be severe political constraints on the casualties the United States will be willing to accept in any future conflict for stakes less than national survival and thus the required relative performance advantage must remain very high.

This assessment assumes that the needed rate of improvement in military systems will slow, but that the Nation will seek to preserve an advantage in key militarily unique technologies to at least match the best of the rest of the world and provide the potential for reduced U.S. casualties and collateral damage in any future conflicts.

This assessment also considers the advanced R&D capability needed to carry out the "prototyping-plus" approach described in chapter 3. One of the findings of this assessment is that R&D must not be pursued, as in the past, according to the "pipeline" model in which research leads to near-term production. Rather, as the present DTIB shrinks and fewer systems are produced, R&D must be pursued with an eye to maintaining superiority in critical technical *capabilities* and as a hedge against technical breakthroughs by potential adversaries, even if the technologies are not incorporated immediately in new weapons. New technology can be demonstrated in laboratories and prototypes, and need not lead to advanced development and production.

Another function of the "advanced" R&D capability needed for the future DTIB is to keep the military community apprised of scientific and technical developments, both military and civilian, throughout the world. Should a global threat arise, the Nation's military establishment would be poised

---

[1] Throughout this report we use common definitions of "research" and "development." "Research" is used to describe investigation intended to advance science and technology without necessarily being directed to a specific end product. The work that follows research and leads to production or specific application is called "development." The Department of Defense (DoD) defines "research" very narrowly to mean only what is generally called "basic research;" for example, what the National Science Foundation classifies as "applied research," the DoD calls "exploratory development." This report's use of "research" corresponds to the activities covered by DoD budget categories 6.1 through 6.3a. "Development" corresponds to budget categories 6.3b and 6.4. Some DoD documents refer to the activities funded under categories 6.1 through 6.3a as "technology base" support but most include only 6.1 and 6.2. The meaning of "technology base" has become muddled so the DoD has created a new category, "science and technology" that clearly includes 6.3a. Thus, figures for "research" cited from DoD documents in this chapter are "technology base" (6.1 and 6.2) funds plus Advanced Technology Development (6.3a) funds.

[2] Weapons performance is just part of the story. As the Gulf War revealed, the quality, training, and organization of the people using the weapons is also an important determinant of the Nation's relative strength.

to exploit the best technology at hand to reconstitute a force that can meet the new challenge. Since a healthy civilian industrial base is an important reservoir of scientific and technical potential, an advanced military R&D capability should also be structured to encourage transfer of technology from the defense sector to the commercial sector.

To best meet future national security requirements, the variety of government laboratories and R&D centers, universities and other nonprofit institutions, and private defense and civilian industrial firms both in the United States and abroad, that carry out today's military R&D will have to change.

The Nation and Congress face fundamental choices regarding the future of defense R&D. What level of effort is appropriate? What should be its scientific and technical focus? Who should perform the R&D and how should it be organized and integrated? What is the proper balance between a near-term and long-term focus? The DoD's ability to influence the Nation's overall R&D base is declining, but its influence is still considerable. U.S. defense R&D, for example, was about 31 percent of all U.S. R&D in 1987 and almost 16 percent of the *total* R&D spending in the European Community, United States, and Japan.[3]

The following sections of this chapter examine how to maintain an advanced R&D capability. The first section reexamines the goals of an R&D effort. The next section addresses the priority of military R&D within the defense budget and the Federal budget as a whole. This is followed by a discussion of technical priorities for defense R&D and problems in the future organization of the defense R&D effort. Finally, the chapter discusses how Congress can affect defense R&D.

MAINTAINING A ROBUST R&D CAPABILITY

A robust R&D capability requires an overall R&D strategy, policies, organization, equipment and facilities, predictable funding, and skilled people to execute the strategy. These requirements are closely intertwined. For example, good people are attracted to research only in part because of salaries. Interesting and meaningful work is at least as important according to many researchers. Front-line research also requires state-of-the-art equipment and facilities. Thus, retaining good people requires meaningful work, good facilities, proper policy, and good management.

Photo credit: Naval Research Laboratory

A radar being tested in 1937 on the roof of the Naval Research Laboratory. Some scientific and technological advances result in revolutionary new military capabilities.

R&D National Security Goals in the New Environment

In the new security environment the United States faces two types of military threats with distinct and characteristic warning times:

1. a currently hypothetical major military threat that—were it to occur—would develop over years, and
2. smaller threats that might flare up with little or no warning.

As a result, defense R&D must have two goals that are not completely complementary:

1. to maintain a scientific and technological capability to guard against surprise and to provide the basis for a buildup—perhaps over several years—of the forces needed to oppose an evolving military threat, and

[3] National Science Foundation, *International Science and Technology Data Update* (NSF-91-309) (Washington, DC: 1991), pp. 5, 9, and 13. These are the most recent published figures for which consistent comparisons between nations are possible.

2. to continually provide standing forces with technology to meet smaller current threats at reasonable cost.

At present the United States is pursuing both paths, supporting R&D and moving to a smaller active military force at a high state of readiness.[4] The reemergence of a major military threat like that of the former Soviet Union (which the administration terms a "reconstituted threat") would require the Nation to build up its forces at least as fast as any potential enemy can. However, until the need for such a buildup arises, the DoD emphasis will shift away from providing current capability toward maintaining potential capability. Current budgets already reflect an understanding that R&D must be maintained. The administration's fiscal year 1993 defense budget request included a 1.5 percent real *increase* in R&D but a 13 percent real *decrease* in procurement.[5]

Tradeoffs between current capability and future potential will affect allocation of resources *within* the defense R&D budget. If current capability is emphasized, a large proportion of R&D would go to development of specific weapon systems, as during the cold war. Emphasis on future capability, however, would shift funding away from development of specific weapons toward more generic research aimed at the development and demonstration of weapon technologies and perhaps to manufacturing technology to produce systems later as the need arises.

The fiscal year 1993 budget request sends mixed signals on the tradeoffs desired by the Administration. The request would cancel or delay several development programs, but development funds are still large and research priorities would not change dramatically. Defense research is proposed to rise from $10.4 billion in fiscal year 1992 to $11.8 billion in fiscal year 1993.[6]

Figure 2-1—Budget Authorization for Research, Development, Test, and Evaluation as a Percentage of Total Defense Authorization

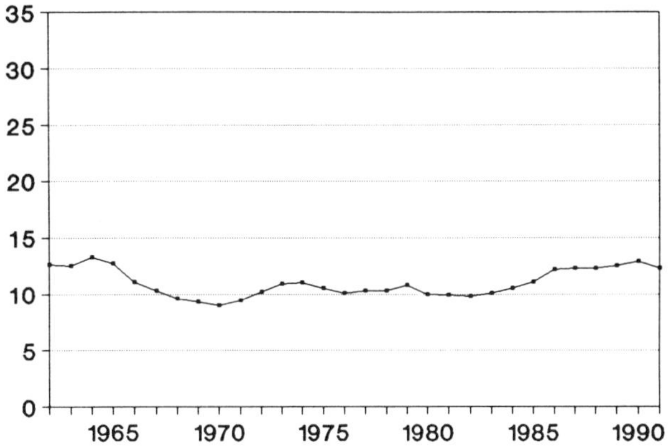

SOURCE: Historical Tables of the Budget of the United States Government.

Funding for R&D

Over most of the last 30 years, spending on R&D, plus the test and evaluation that is a part of any development program (i.e., RDT&E), has been about 10 to 11 percent of the DoD budget.[7] (See figure 2-1.) If historical ratios continue, absolute funding for defense R&D will shrink along with the rest of the defense budget, unless there is a commitment to support R&D as a means of maintaining military potential.[8] The threats the Nation faces have not been reduced evenly across the board and there remains the possibility that a major military threat could emerge. Thus, even if a large reduction in current capability is warranted, it does not necessarily follow that investment in military R&D should be reduced proportionately.

Most independent projections are, however, that the resources for R&D will decline in the future. For example, the Electronic Industries Association (EIA) annually makes 10-year projections of defense spending that have been accurate in the past. The

[4] Military planners appear to give a high priority to the training and readiness of whatever active force the Nation keeps. This may be reflected in press speculation about future shifts of resources from procurement to training. See, Andy Paztor, "Pentagon Weighs Cut of $50 Billion," *The Wall Street Journal*, Dec. 20, 1991, p. A3.

[5] Dick Cheney, *Annual Report to the President and the Congress* (Washington, DC: U.S. Government Printing Office, February 1992), p. 131.

[6] Department of Defense, *RDT&E Programs (R-1)*, Jan. 29, 1992, p. III. These figures are the sum of the "Technology Base," that is, budget categories 6.1 and 6.2 and the "Advanced Technology Development," that is, 6.3a.

[7] Executive Office of the President, Budget of the United States Government, Fiscal Year 1992, Part 7, Table 3.2. Test and evaluation adds about 25 percent to the R&D budget.

[8] Extensive prototyping, including limited production of weapons for operational testing, may be expensive compared to current development programs. Thus, a prototyping effort could dominate future R&D budgets if budgeted under development rather than production.

Figure 2-2—Nominal Decline in Future Spending for Research, Development, Test, and Evaluation

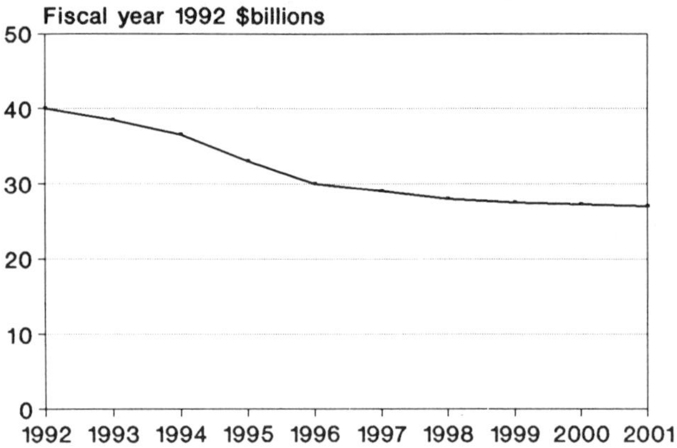

SOURCE: Electronic Industries Association

EIA forecast is based on an assumption that RDT&E's fraction of the budget will grow only very slightly, to 12.9 percent in 2001. Thus, their predicted defense budget of $208 billion in 2001 implies a decline from $40 billion today in total RDT&E spending to $27 billion in 2001 (all in fiscal year 1992 dollars). (See figure 2-2.)

The DoD's own plans also anticipate reduction in R&D over the next few years. Air Force RDT&E budgets will fall most steeply, from $15 billion in 1992 to $8 billion in 1997.[9] Current plans call for the Army's RDT&E budget to decline slowly from $6 billion in 1992 to $4 billion in 1997.[10] These budget projections should be viewed cautiously because changes in development funding for just a very few very expensive projects can skew the entire budget.

Maintaining an advanced R&D capability may be relatively more expensive for the DoD in the future, since, with less production, firms will have little incentive to share in paying for R&D costs. But justifying a particular level of R&D spending is difficult. For generic research, the problem is to show a clear relationship between the effort and the result.[11] Each Service terms its research a "corporate investment," i.e., a cost of maintaining expertise. Research support requests for fiscal year 1993 are $1.8 billion for the Air Force, $1.5 billion for the Navy, $1.1 billion for the Army, and $7.2 billion for DoD agencies.[12] Private companies determine levels of research spending on the basis of investment that they believe is needed to maintain a competitive edge over rivals. The Nation's military establishment should similarly monitor technology developments of other countries to determine the research needed to maintain the desired relative advantage in weapon performance.

The criteria for deciding development funding levels are changing. In the past, the Nation could decide on the needed rate of introduction of new weapons to maintain qualitative superiority over the Soviet Union. But in the future, to respond to a more ambiguous array of threats, the Nation will need to maintain a range of industrial, technical, and manufacturing skills, possibly through a prototyping-plus strategy, a big part of which would probably show up in the development budget. (See ch. 3.)

Allocation of Funds Among R&D Performers

Reduced defense R&D spending will change the *distribution* of R&D effort among industry, university, nonprofit, and in-house Service laboratories. The current distribution for research is shown in figure 2-3. Unless offsetting actions are taken, reduction in defense research funding will cause a relative increase in research activity by Service laboratories and a decrease by private industry. There will be less industry incentive to support research, while government laboratory managers may try to keep a relatively larger slice of a shrinking pie in-house.

The trends for distribution of development funding are less clear. Over two-thirds of development now occurs in industry. Total development funding may stay high if the Nation pursues a prototyping-plus strategy like that described in the next chapter. This activity would most likely remain in industry. Many companies argue, however, that current ap-

[9] Congressional Budget Office, Staff Memorandum, "The Costs of the Administration's Plan for the Air Force Through the Year 2010," December 1991, p. 18. Some analysts argue that Air Force R&D figures are suspect because they may include large secret production programs.

[10] Congressional Budget Office, Staff Memorandum, "The Costs of the Administration's Plan for the Army Through the Year 2010," December 1991, p. 21.

[11] See Genevieve J. Knezo, *Defense Basic Research Priorities: Funding and Policy Issues*, Congressional Research Service Report for Congress, Oct. 24, 1990, for a discussion of the problems of determining funding levels and setting research goals.

[12] *Defense Daily, Special Supplement, Details of 1993 DoD Budget Request*, Jan. 31, 1992, pp. S-1 to S-3. Note that the research budget for the defense agencies is dominated by the Strategic Defense Initiative Organization.

Figure 2-3—Possible Distributions of Future Research Effort

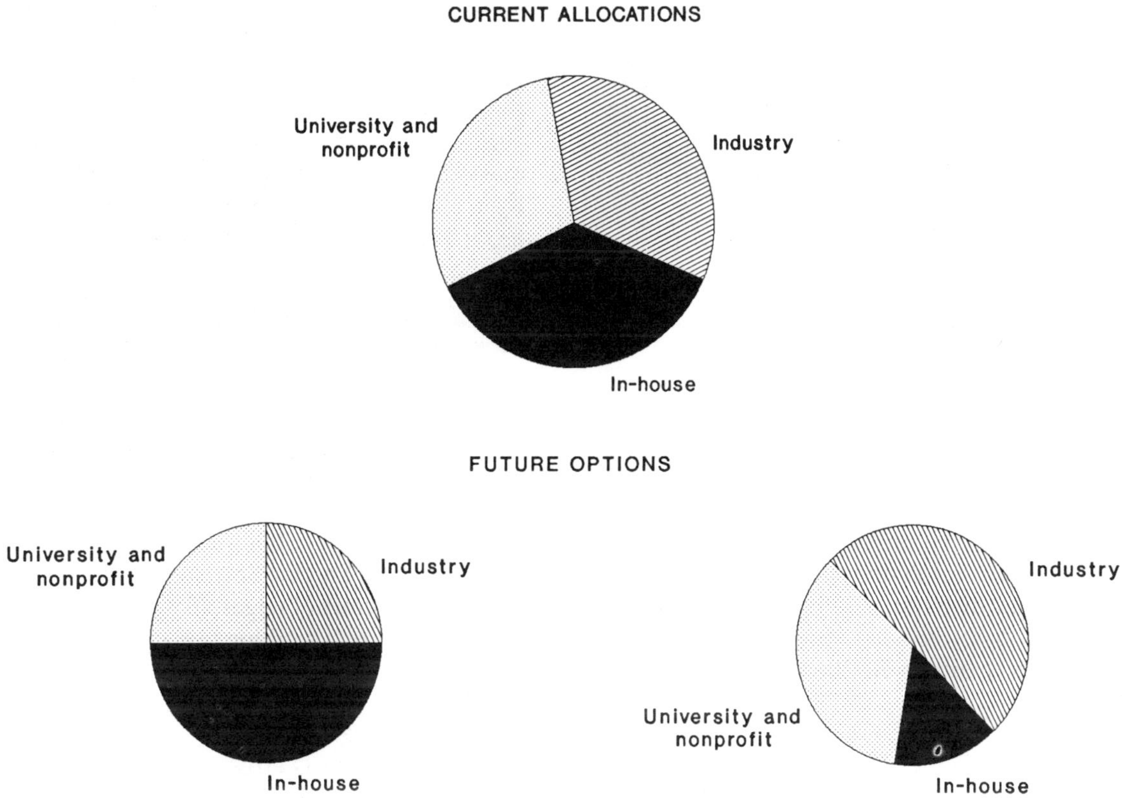

Option 1: With emphasis on preserving militarily unique technology in government laboratories

Option 2: With emphasis on dual-use and commercial technology

SOURCE: National Science Foundation, *Federal R&D Funding by Budget Function,* and OTA.

proaches to prototyping are not commercially viable. Unless the DoD takes steps to make prototyping profitable for industry, design and prototyping activities might need to move into government laboratories or arsenals.

OTA's defense industry survey indicated that most companies, foreseeing reductions in production contracts, are planning to cut their own spending for R&D. "Independent research and development," or IR&D, is a company's R&D that is funded outside of explicit government R&D contracts.[13] Over the long term, the IR&D that can be recovered from production contracts as overhead will decline as procurement declines. Action has already been taken to counter some of these trends. Companies' allowed IR&D recovery rate was increased under legislation passed last year, for example. But direct DoD funding to industry may need to increase in critical R&D areas to maintain current levels of effort.

Any decline in government-funded R&D will exacerbate reductions in company-financed R&D. Most studies indicate a positive correlation between federally-funded R&D in a company and the company's own R&D expenditures.[14] This finding indicates a leveraging of Federal funding: a dollar of

[13] There is some confusion in the literature about the definition of "IR&D." The formal government definition is any noncontract-funded R&D. Thus, for example, all of the R&D of a company with no connection to the DoD at all is considered IR&D. (See Office of the Director of Defense Research and Engineering, *The Independent Research and Development Program, A Review of IR&D,* June 1974.) Many writers commonly use the term to refer to only that portion of a defense contractor's R&D that is potentially recoverable as an allowable overhead expense on a government contract. This section will refer to a company's overall effort as IR&D and the recoverable portion as recoverable IR&D if there is an ambiguity.

[14] For a survey of results, see Frank R. Lichtenberg, "The Effect of Government Funding on Private Industrial Research and Development: A Reassessment," *Journal of Industrial Economics,* vol. 36, No. 1, September 1987, pp. 97-104.

Photo credit: Pratt and Whitney

New technologies are first demonstrated and proven in this core engine before the manufacturer applies them to a new or existing engine design. Most of the Nation's defense technical and expertise resides in companies.

Photo credit: Wright Laboratories

Research on new materials is conducted by the Air Force at Wright Laboratories. The Service laboratories are important centers of research and pride themselves on being closely tuned to the needs of the military users.

Federal funding creates more than a dollar of total R&D in both defense and nondefense sectors. Presumably therefore, as Federal defense R&D support shrinks, overall R&D shrinks even more.

Service laboratories have built up important areas of expertise and are well tuned to Service military requirements. But critics of reliance on Service laboratories for DoD R&D note that they work for a single Service, and sometimes only one element within that Service, therefore their view may be too narrow and they may overlook technologies that do not promote current Service missions. In the extreme, this is essentially a legal restriction; for example, the Army cannot own, hence does not support R&D in, fixed-wing combat aircraft. Furthermore, civil service regulations are said to stifle research creativity in government laboratories. Critics of Service laboratories argue that private companies, universities, and other outside research organizations are likelier to think up a potential mission to illustrate a need for some new technology that they have just developed. Salesmanship usefully gets new ideas into consideration.

Service R&D activities are coordinated through the Office of the Secretary of Defense (OSD), but the Service laboratories also have their own group, the Joint Directors of Laboratories (JDL) under the auspices of the Service Secretaries. The JDL is made up of the directors of research or laboratories of each of the Services. In addition to meeting as a group, the JDL sets up various subpanels of Laboratory Technical Directors or Chief Scientists, to discuss and coordinate work in particular areas.[15]

Under the auspices of JDL, the Services are undertaking a major change in their approach to allocating research effort among themselves, under a directive from OSD to coordinate their technology support. The result is a program called "Tri-Service S&T Reliance" (previously called Project Reliance). The objective of this program is to minimize redundancy among Service laboratories. At the very least, similar efforts at different laboratories should be coordinated. Where appropriate, efforts will be physically consolidated at a single laboratory and one Service will be designated a lead Service. For example, the Navy recognizes the Army's extensive expertise in large-caliber guns, so all Services' relevant gun technology development will take place at the Army's Picatinny Arsenal. Similarly, fuel and lubricant research will take place at the Air Force's Wright Laboratory, while work on space-based infrared sensors will be at the Naval Research Laboratory.[16]

[15] Fredrick R. Riddell et al., *Report of the Task Force for Improved Coordination of the DoD Science and Technology Program* (Alexandria, VA: Institute for Defense Analyses, August 1988) vol. II, app. C.

[16] Joint Directors of Laboratories, *White Paper on Tri-Service Reliance in Science and Technology*, Office of Naval Technology, January 1992.

Reliance is still in its early stages so it is too soon to evaluate results.[17] If current plans are carried through, however, it could bring a fundamental change in how the Services organize their technology research.

A robust civilian industrial base is important to the DTIB. If the defense R&D effort is given the additional goal of helping civilian technology, R&D funded through the DoD could expand substantially. There is no consensus, however, on the degree to which the Federal Government should support civilian industrial R&D, nor on whether such support should come from the DoD.

Some advocates of direct government support of civilian industrial R&D argue that while such support is essential, the regulatory barriers that government has erected between military and civilian enterprise are so great that channeling the money through the DoD is extremely inefficient. Other advocates concede the inefficiency, but counter that the government has no current mechanism with adequate scope and experience other than DoD to mount such a program. Alternative programs supported, say, through the Department of Commerce, might take years to build up. Further, they argue that the political realities are that cuts in DoD R&D are not going to be transferred to some other research agency, such as the National Science Foundation or the Department of Commerce's National Institute of Standards and Technology (NIST). Thus, according to this argument, the alternative to supporting civil industry inefficiently through the DoD is not to do it at all.[18]

In fact, the alternatives are not so stark. Other government programs, while currently small, could be expanded. For example, the Advanced Technology Program (ATP) under NIST is growing, with a $50 million fiscal year 1992 budget and an administration request of $68 million for fiscal year 1993 and much support in Congress. OTA's assessment of industrial competitiveness, *Making Things Better*, argues through analogy with the Defense Advanced Research Projects Agency (DARPA) that the ATP could use effectively over $1 billion per year.[19]

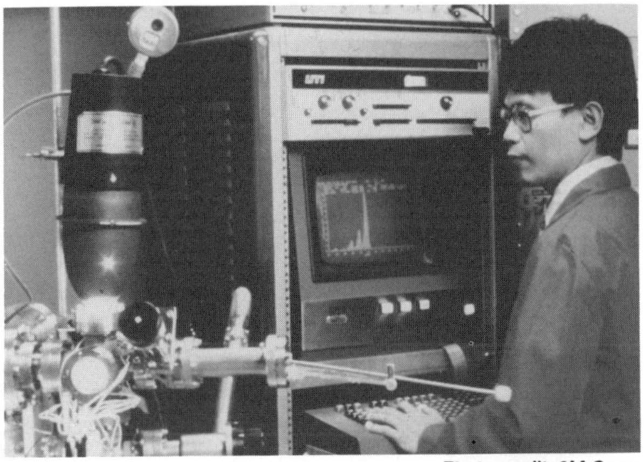

Photo credit: 3M Corp.

R&D in commercial laboratories is increasingly sophisticated and will be important for future military technology.

Proponents of greater civil-military integration argue that reducing bureaucratic barriers between military and civil industry would greatly increase the number of sources of new technology. Using more integrated commercial firms, however, will complicate the issue of foreign dependence since many of the large and most innovative companies are multinationals. These companies also offer, of course, ready access to valuable technology abroad. To make best use of civilian technology, changes in DoD procurement and contracting practice are required, as discussed below.

The Nation's universities have traditionally been strong in long-term basic research. Although basic research in universities is small in dollar terms when compared to the DoD budget, it is the primary source of fundamental scientific advances and, just as importantly, to the training of future scientists and engineers.

The Department of Energy operates the National Laboratories that are responsible for the development and testing of nuclear weapons. With the end of the cold war, further advances in nuclear weapons are far less urgent but nuclear weapon design know-how is something the Nation cannot afford to lose.

---

[17] Implementing letters were signed only in September through November of 1991.

[18] There is little agreement on the extent or net benefit of military R&D on the civilian economy. A recent report containing a review of the important existing literature is, David Gold, *The Impact of Defense Spending on Investment, Productivity, and Growth* (Washington, DC: The Defense Budget Project, 1990).

[19] See U.S. Congress, Office of Technology Assessment, *Making Things Better: Competing in Manufacturing*, OTA-ITE-443 (Washington, DC: U.S. Government Printing Office, February 1990), p. 76.

Photo credit: MIT Lincoln Laboratories

This university laboratory developed thin films of superconducting materials. The Nation depends on universities for much of its long-term basic research.

To summarize, industry holds most of the defense technology and almost all the knowledge about manufacturing. Further, the increased importance of dual-use technology could make industry a better potential provider in the future. At present, however, it seems likely that industry will lose much of its incentive to maintain supporting defense technology. The DoD needs either to make the R&D profitable in its own right or become a truly innovative in-house technology developer. In the latter case, government laboratories and arsenals would have to take on more technology development, but they would need to be careful to maintain communication with potential manufacturers to ensure that designs are producible. The government should carefully consider maintaining current DoD support for university R&D and explore ways to utilize the civilian sector.

Setting R&D Technology Priorities in the New Security Environment

The new security environment is changing the relative importance of many military missions. Sometimes the technology implications of new mission emphases are fairly clear. For example, any Navy shift of emphasis from open ocean to shallow water operations will require more attention to countering mines. Similarly, if the Army is less concerned with building heavy tanks for war in central Europe and more with deploying armored forces to unpredictable trouble spots, its R&D emphasis should shift to making weapons lighter and easier to maintain under diverse field conditions.

There is a requirement to allocate R&D funds across technology areas. Congress is, of course, concerned about whether the allocation is correct but is not well-suited for setting detailed R&D goals with the current approach. Congress and the DoD set overall military missions and review needed technology developments but without complete coordination between these two processes.[20]

In the absence of any published DoD technology plan, Congress requested that the Office of the Secretary of Defense (OSD) prepare a "critical technology plan."[21] The resulting report continues to be criticized as providing merely a list of technologies rather than an investment plan that explains how to apply these technologies to military missions. Some critics argue that the list is of limited value in allocating resources because the technology areas are so broadly defined that very little is not considered "critical." They argue, furthermore, that this generality is a consequence of fears that the list will be used in a simple-minded way for funding decisions: if work is not labeled "critical" it will get cut. Others argue that the critical technology plan concentrates only on weapon technology and overlooks training or logistics, which are just as critical to military strength.

The individual Services do have technology investment strategies that are coordinated through OSD. The Army's Technology Base Master Plan,[22] for example, lays out an investment strategy for implementing technology goals, an explanation of how to get from here to there, and how much it will cost. The OSD has recently developed several Technology Thrust Areas that should make clearer

[20] See U.S. Congress, Office of Technology Assessment, Background Paper, *Evaluating Defense Department Research*, June 1990.

[21] P.L. 101-189 103 Stat. 1512 Paragraph 2508(a) "Annual Plan.— (1) The Secretary of Defense shall submit to the Committees on Armed Services of the Senate and House of Representatives an annual plan for developing the technologies considered by the Secretary of Defense and the Secretary of Energy to be the technologies most critical to ensuring the long-term qualitative superiority of United States weapon systems. The number of such technologies identified in any plan may not exceed 20. Each such plan shall be developed in consultation with the Secretary of Energy."

[22] Deputy Assistant Secretary for Research and Technology, *Army Technology Base Master Plan*, vols. I-III (Washington, DC: Department of the Army, November 1990).

Photo credit: DoD

This tank is only lightly armored. It is more vulnerable than a heavy tank but easy to transport by air. As the military threats facing the Nation change, the technical goals of defense R&D must be adjusted.

the route from the DoD's overall technology investment to its military missions. Currently, there are seven thrust areas—surveillance, precision strike, air superiority, antisubmarine warfare, land combat vehicles, readiness and training, and affordability[23]—but the areas could change in the future in response to new security or technical developments.

With slower deployment of new major weapon platforms, relatively more attention will be given to subsystems and upgrades of the platforms and the munitions that they carry. The development budget may include prototypes to demonstrate new technology or designs for which there is no immediate plan for actual development for quantity production. With falling budgets, some Service R&D planners see greater research emphasis on reliability, durability, and efficiency as a way to reduce operating costs. The increasing complexity and decreasing number of weapon systems have led the DoD to emphasize designing for producibility and manufacturing process.

The changes in military requirements will have less effect on priorities at the research end of the R&D spectrum. For example, no matter what the threat, there is a strong consensus that R&D will continue or even increase its current emphasis on "information technology," including sensors, data analysis, and displays, along with communications, computers, software, and storage and manipulation of data for manufacturing. However, even with a large shift in emphasis toward, for example, electronics, the basic goals of electronics research will remain much the same: reduced size, lower power requirements, higher speed, lower cost, and greater reliability, whether the end-result eventually appears in a ballistic missile or a shoulder-fired rocket. Similar arguments will apply for a range of technologies, from biotechnology to materials.

The general trend since the end of World War II has been for Congress and the Executive Branch to require ever clearer justifications of military R&D, usually in terms of final military application. This goal is most clearly stated in the Mansfield Amendment, P.L. 91-441 Title II Section 204:

> None of the funds authorized to be appropriated to the Department of Defense by this or any other Act may be used to finance any research project or study unless such project or study has, in the opinion of the Secretary of Defense, a potential relationship to a military function or operation.[24]

Although this requirement, if broadly defined, should be easy to meet, many R&D managers argue that it has had the effect of biasing project selection toward those whose military connection is not just "probable" but most obvious, with the effect of narrowing the defense R&D base.

As the Nation broadens its definition of "national security" to include international economic competitiveness, the past emphasis on the narrow military justifications for defense R&D will be challenged. Many observers are concerned about the Nation's industrial performance and view inadequate investment in civilian R&D as part of the problem. Since military R&D is almost one-third of the Nation's total public and private R&D expenditure—a higher fraction than that of other western industrial countries[25]—one approach is to tap into military R&D to help civilian enterprise. (See figure 2-4.) This

[23] Briefing from the Office of the Deputy Director, Research and Engineering (Plans and Resources), Feb. 9, 1992.

[24] This wording superseded the even stronger wording enacted the previous year that projects should have a " . . . direct and apparent relationship to a specific military function . . . " (P.L. 91-121). The wording was altered again slightly in P.L. 100-370 to allow the DoD to spend R&D funds, "for purposes related to research and development for which expenditures are specifically authorized in other appropriations of the Department of Defense." This recent change indicates a reversal in the trend toward narrow military justification toward a concern with a broader national security.

[25] National Science Foundation, *International Science and Technology Data Update: 1991*, NSF 91-309 (Washington, DC: 1991), pp. 10-15.

Figure 2-4—Defense R&D as a Percentage of Total Government R&D Spending and as a Percentage of Total National R&D Spending

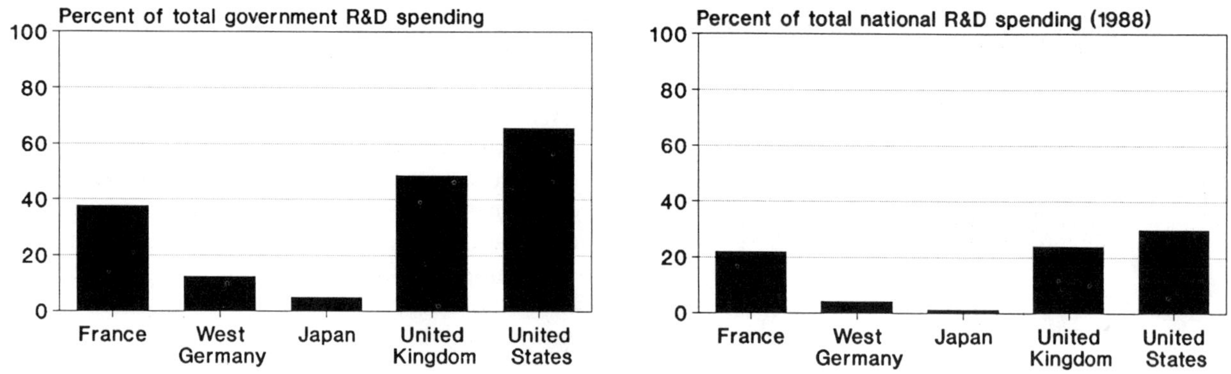

SOURCE: National Science Foundation, *International Science and Technology Data Update: 1991.*

approach may include adding civilian industrial competitiveness to the criteria by which military R&D proposals are judged. Congress frequently has the dual objectives of military and commercial benefit in mind when it supports ostensibly military projects directed at improving manufacturing, such as the MANTECH and SEMATECH programs described in *Redesigning Defense*.[26]

The DoD and the administration resist this approach because of the difficulty of balancing civil and military objectives. Defense managers, while aware of the importance of industrial competitiveness, are hesitant to use it as a criterion for funding military R&D. Their general concern is that resources devoted to important military needs are already limited, if not inadequate, and that additional nonmilitary objectives would make fulfilling those needs even more difficult. Changes in the bureaucratic incentives for integration of civil and military R&D will have to be made at higher levels, i.e., Congress or the President, before managers adopt such criteria.

Another approach would be to keep current military R&D priorities and improve the transfer of military technology to the private sector. Whether government support for certain industrial technical development projects is warranted has been a subject of much debate. Until that debate is resolved, however, the question remains whether industrial development funds should be funneled through the DoD. As described below, the substantial barriers between defense and commercial business sectors resulting from the special legislative and regulatory environments created by the Federal Government hamper the transfer of technology.

A third approach would be to maintain current defense R&D priorities but reduce the overall level of R&D funding funneled through the DoD. Released funds could support commercial R&D directly through some civilian government organization, or the funds could go to indirect support, for example, tax incentives for R&D. This approach would compel the DoD to obtain more of its technology from the civil sector and adapt its doctrine to suit available technology. R&D might also focus on strategic economic vulnerabilities that pose a threat to national security. For example, if an extensive R&D program had produced energy independence for the West, Iraq's invasion of Kuwait would have had a fraction of its actual importance.

Organization of Government Support of Defense R&D

The organization of R&D must balance the needs of the operational military community, the R&D community, and the Nation's defense effort as a whole.[27] How to maintain this balance will be important in the years ahead as the DTIB shrinks.

[26] MANTECH is a program to support MANufacturing TECHnology and SEMATECH a government-industry consortium to develop SEmiconductor MAnufacturing TECHnology. See *Redesigning Defense*, pp. 52-54.

[27] See U.S. Congress, Office of Technology Assessment, *Holding the Edge: Maintaining the Defense Technology Base*, OTA-ISC-420 (Washington, DC: U.S. Government Printing Office, April 1989) for a fuller discussion of DoD R&D management and organization.

This issue is of direct concern to Congress, since Congress has historically been involved in the organization of DoD R&D.

If the primary objective is relevance of R&D to the users' perceived immediate needs, then R&D should be closely tied to the acquisition function, which in turn should be under the control of the Services. The danger with this arrangement is that it might focus on short-term problems. If the objective is to emphasize long-term technological support, the earlier science and technology work could be given more visibility, perhaps by having the person in charge of R&D report to the secretary level in OSD or the Services, rather than through the person in charge of acquisition. Another issue of concern to Congress is that, in a declining budget environment, existing R&D groups may resist consolidation or redirection.

Present Structure and Consolidation Plan

Until 1986, the Director of Defense Research and Engineering reported directly to the Secretary of Defense. In response to the widespread perception that the acquisition process was not adequately managed, the Goldwater-Nichols reorganization of that year created an Under Secretary for Acquisition, to whom the DDR&E reported. This reorganization may have increased the communication between the acquisition and the R&D communities, but it also reduced the visibility of the R&D issues to the Secretary.

Assistant Service Secretaries in charge of R&D report through their Service chains of command, but also coordinate through OSD. Some critics contend that coordination is insufficient and more centralized control is needed. Advocates of increased centralized control argue that redundancy and inefficiency are inevitable if each Service handles its R&D separately. While R&D redundancy was desirable in an era of growing or level budgets, declining budgets should force greater coordination.

In contrast, Service R&D managers argue that independence from OSD is vital because the Services best understand the needs of the ultimate users of the technology. Moreover, rivalry among the Services produces alternatives that might not have surfaced if the research agenda were centrally controlled. A good example is the development of the Navy's Polaris submarine at a time when the strategic nuclear role was dominated by the Air Force. If there had been a central strategic nuclear R&D directorate at the time of Polaris' proposal, it probably would have been dominated by advocates of ICBMs and bombers and the submarine-launched ballistic missile, which has become the cornerstone of the U.S. nuclear deterrent, might never have been pursued.[28]

Each of the Services is reorganizing or planning to consolidate its laboratories. Service laboratories and research centers perform in-house R&D and administer projects contracted to private industry. There are 66 Service laboratories (76 if the laboratories making up the Air Force "superlabs" are counted individually). In 1990, the laboratories employed 60,000 people of whom 26,000 were scientists and engineers. Total funding was $6.5 billion, over half of which went to externally performed contract R&D with part of the remainder going to management of these outside contracts.[29] The number of employees has shrunk somewhat since.

The Army has the most extensive reorganization plan, resulting from its "Lab 21" study. It plans to consolidate most administrative control and many activities in a single Combat Materiel Research Laboratory in Maryland. Currently, similar technologies often are explored in different Army centers if they have different end uses. Some of these centers would be consolidated under the plan. For example, combat vehicle propulsion research is based in Warren, MI, the home of the Tank and Automotive Command, while aviation propulsion research is based in Cleveland, OH, on the site of NASA's Lewis Research Center. The Army believes that, at the research level, these two applications are similar enough to warrant future consolidation of the laboratories at Cleveland.[30]

[28] James R. Schlesinger, *Defense Planning and Budgeting: The Issue of Centralized Control* (Washington, DC: Industrial College of the Armed Forces, 1968), p. 18.

[29] Federal Advisory Commission on Consolidation and Conversion of Defense Research and Development Laboratories, *Report to the Secretary of Defense*, September 1991, p. 1.

[30] See George Singley, testimony before the Subcommittee on Defense Industry and Technology, Committee on Armed Services, U.S. Senate, May 21, 1991.

The Navy is consolidating its R&D activities into four existing Research, Development, Test and Evaluation Engineering and Fleet Support Centers, and one Service-wide research laboratory. Unlike the Army's technology-center approach, the Navy's centers are organized around war-fighting missions. For example, Navy R&D relating to air warfare will be headquartered at the Naval Air Systems Command, with a weapons division at Point Mugu and an aircraft division at Patuxent River. The other centers are the Naval Surface Warfare Center (which will include surface-based antisubmarine warfare), the Naval Undersea Systems Center, and the Naval Command, Control, and Ocean Surveillance Systems Center (which covers surveillance and communications). The Naval Research Laboratory, which explores a broad range of research areas, will remain as a Navy-wide laboratory under the Office of Naval Research.

The Air Force reorganization plan includes the consolidation of 14 laboratories into 4 "superlabs."[31] However, no laboratories are planned to close or move immediately, although management and administrative functions will be concentrated in the four central laboratories.

The Federal Advisory Commission on Laboratory Consolidation endorsed the Service consolidation plans almost without exception. According to the Commission report:

> The laboratories provide to the acquisition agents (i.e., the Services' program managers), an in-house, technologically qualified agent to oversee or evaluate the performance of the industrial developer as required to ensure that the design is technically sound, will satisfy performance requirements, and is producible and affordable.[32]

The possibility of laboratory mission changes and their implications for organization and size are not developed in the Commission's report.

Medical laboratories are examples of *cross-Service* consolidation. There is no reason in principle that other technologies in addition to medicine could not be similarly coalesced, as the Federal Advisory Commission suggests considering for microelectronics.

A number of alternatives for further consolidation exist. OSD could, for example, consolidate research activities while leaving development to the Services, or OSD could, in the extreme, assume control of all Service R&D activities following the French R&D and acquisition model. The Services argue against consolidation on the grounds that R&D will become disconnected from their direct needs. And they argue that as long as the Services are responsible for acquisition, they should be responsible for the supporting R&D. But in a future circumstance of declining budgets and continuing need for technological advance, consolidation *across* Services may be as necessary as consolidation *within* each Service. There are also certain joint tasks that OSD might best oversee among the Services, such as communication, data fusion and dissemination, and attack coordination.

The OSD does not operate any laboratories, although it has two Federally Funded Research and Development Centers (FFRDCs): the Institute for Defense Analyses (with $96 million in fiscal year 1990) and the Logistics Management Institute (with $21 million in fiscal year 1990[33]). OSD also supports defense agencies like the Defense Nuclear Agency, the Defense Advanced Research Projects Agency, and the Strategic Defense Initiative Organization. While none of these are laboratories, they have resources to support R&D at Service laboratories or elsewhere. Total R&D funding of the organizations funded through OSD was $6.9 billion in fiscal year 1991.[34]

Issues for the Future

Congress might consider changing the balance between Service autonomy and OSD coordination of R&D. To reduce redundancy, it could funnel all R&D funds up to some level (perhaps 6.3a) through OSD, perhaps by extending the model of inter-Service medical laboratories to other areas. Alternatively, Congress could encourage the Services to continue the approach begun with Reliance, that is, to assign responsibility for each important technol-

[31] *Report to the Secretary of Defense*, op. cit., footnote 29, p. 18.

[32] *Report to the Secretary of Defense*, op. cit., footnote 29, p. C-1.

[33] Dollar figures from Michael Davey, *Defense Laboratories: Proposals for Closure and Consolidation*, Congressional Research Service, 91-135 SPR, Jan. 24, 1991, p. 9.

[34] Department of Defense, *RDT&E Programs (R-1)*, February 1991, p. D-II.

ogy area to a single Service, which would support the other Services in that area.

As the DTIB gets smaller, Congress will want to have military R&D activities organized such that no important R&D mission is overlooked. For this, Congress will need a better idea of how technologies relate to military missions and how the entire defense R&D effort is coordinated. Ideally, the system should be designed to foster new ideas and avoid parochialism, so coordination does not become stifling overmanagement. Achieving this state should be a key goal of R&D reorganization.

OTHER CRITICAL ISSUES

Congress will need to address a number of specific critical issues relating to the organization and function of the defense R&D base.

Defense R&D Personnel: Maintaining the Know-How

A critical objective of defense R&D policy in the new era is to maintain the core skills and knowledge that are key to the whole defense enterprise. Personnel reductions in defense R&D must be carried out carefully to retain key skills, and those that exist only in the defense base must be maintained there. Some knowledge and capability exists in groups of people rather than individuals. These groups may require special treatment if their skills are not to be lost. For example, a prototyping-plus strategy, discussed in the next chapter, can help maintain critical pools of design and development talent.

Many of the concerns about government laboratory personnel long preceded the ongoing reduction in the size of the DTIB. Some problems may be exacerbated by future shrinkage, while others may be made more manageable. For example, the question of salary seems to be permanent.[35] Government pay for scientists and engineers lags behind that of comparable positions in industry. Measuring the lure of intangibles such as job security is, however, hard and thus predicting the exact effect of their loss is also difficult.

Some argue that salaries for scientists and engineers in private-sector defense firms are inflated by up to 15 percent relative to comparable nondefense sectors and, moreover, that this difference has drained the Nation's broader civilian industrial base of its best technical talent.[36] As international commercial competitiveness increases in importance relative to defense, the Nation may have less interest in maintaining whatever defense salary bonus might exist, and want to encourage good people to work in the civil sector.

If the mission of the Service laboratories changes—for example, shifting emphasis from oversight of contract R&D toward more direct involvement in R&D work—then the personnel requirements also change. If the Service laboratories increase their role as developers of technology, then the quality of their personnel may also need to improve. This may require further changes in pay scale. Just as important are changes in "revolving door" policies that inhibit movement of personnel between government and industry. If the primary function of a government scientist is to be an adviser to a government buyer, then there is a need to forestall conflict of interest by maintaining a clean separation between the scientist and industry. If the role of government scientists is to develop new technology and promote technology transfer, then scientists should be positively encouraged to move back and forth between government laboratories and leaders in industrial technology.

The DoD is an important source of support of research in universities.[37] In electrical engineering, for example, the DoD provides the majority of university research support. In some other fields (e.g., aeronautical and mechanical engineering), the DoD is the largest single source of funds. Funds from the DoD support much research critical to the Nation's military capabilities, but another important function of DoD research funding for universities is the training of students who then enter the Nation's pool of scientific and engineering talent. A reduction in DoD funds for university research is possible as overall defense budgets shrink. Congress or the OSD may want to maintain funding of university research

[35] Bureau of the Budget, *Report to the President on Government Contracting for Research and Development*, May 1962, Annex 5, pp. 47-49.

[36] Joel Yudken and Ann Markusen, Rutgers University, "The Labor Economics of Conversion: Prospects for Military-Dependent Engineers and Scientists," p. 15, a paper presented at the conference, "Engineers and Economic Conversions," University of Arizona, Tucson, July 15-17, 1991.

[37] Knezo, op. cit., footnote 14, p. 24.

to continue to have access to this important source of technological strength.

As the DTIB shrinks, the need for scientists, engineers, and technicians will also decline. Meeting the lesser personnel demand should, in the long term at least, be an easier task than attracting adequate numbers was in the past. During the transition to a smaller base, however, the major danger is that past investment that has resulted in a huge reservoir of experience and knowledge, both in individuals and in groups, will be lost. Moreover, the smaller year-to-year demands of the DTIB for technical talent will not reflect the Nation's requirements if a major new military threat arises. Thus, the Nation should be mindful of where the DTIB technical talent goes as it leaves the base. There is a difference, both to national security and national prosperity, between moving scientific and technical talent between defense and civil work and not being able to use it in the economy at all. Industry will continue to supply the great majority of R&D personnel important to defense, which can increasingly include R&D talent outside of the traditional defense companies if civil-military barriers are reduced.

Independent Research and Development

Companies that have contracts with the U.S. Government negotiated on the basis of their costs, rather than market or bid prices, are allowed to charge as an overhead expense the "normal" costs of doing business. Since before the Second World War, the government has considered limited R&D and other engineering efforts as allowable overhead costs.

In the past, the IR&D recovery scheme exacerbated the separation of military and civilian technology. If military and commercial business, including R&D, is mixed in one company division, then that portion of R&D judged to have a potential interest to the DoD must be prorated between the government and commercial business. If the optimal R&D investment in the government and commercial parts is not the same, then anomalies result. For example, if the military products warrant a higher rate of R&D investment than the commercial products, and IR&D recovery rules require prorating R&D costs, then the company's commercial products will be more expensive than those of a competitor that does no military R&D. Thus a company that does both commercial and military production and R&D has yet another incentive to separate its two customer lines, creating yet another barrier between commercial and military technology transfer.

Today the trend is toward encouraging civil application of recoverable IR&D. To qualify for recovery as an overhead cost on a defense contract, R&D must now be shown to be of "potential interest" to the DoD.[38] The law states that IR&D regulations should encourage contractors to pursue R&D activities that

1. strengthen the DTIB,
2. enhance industrial competitiveness,
3. promote critical military technologies,
4. develop dual-use technology, and
5. develop technology to benefit the environment.

A broadening definition of what is of interest to the DoD combined with higher recovery rates does not give companies a blank check to charge R&D to government contracts. Contract officers must still agree that the charges are reasonable.

Substantial additional changes in IR&D recovery rules may be needed to change the way companies support R&D. Recovery of IR&D as an overhead expense on procurement links R&D to production. This linkage will not be desirable in the emerging era of production cuts. For example, production of many types of weapons may fall sharply during the transition to a smaller military, while the need for R&D will remain high. In these instances, contract R&D could make up for reduced overhead recovery of IR&D expenses. In addition, if the DoD buys commercial technology incorporated into military-specific products, then companies will want some simple mechanism for folding past R&D costs into the price of the products.

Technical Data Rights

When the DoD acquires a product, it typically acquires some license right to the "technical data" related to that product.[39] Technical data could be just "form, fit, and function" information, that is, a

[38] Independent Research and Development: CFR, Title 48, Chap 2, sec. 231.205-18 and P.L. 101-510, sec. 824. P.L. 101-510 put "interest" in place of the earlier requirement for a "relationship" and significantly broadened what is of "interest" to the DoD to include international competitiveness.

[39] CFR, Title 48, ch. 1, sec. 52.227-14.

description of what a device does and where the holes for bolting it in need to be drilled. At the other extreme, the government sometimes demands the right to manufacturing data detailed enough that another firm can produce the item. The government argues that it has vital national security responsibilities that transcend normal market requirements and justify extensive data requirements. But many firms are concerned that technical data rights rules cause them to lose commercially useful technology.

Current policies toward technical license rights inhibit mixing of military and commercial R&D. If the government wants to assure a second source for some critical item, it may request only government-purpose license rights. This is technical data that allows the government or another of its contractors to build the item, but any contractor that does so must not use the data for any commercial purpose. No one can guarantee that such separation will be effective in all cases. Thus, companies that have technology of commercial interest are reluctant to sell products to the government along with license rights as they fear that the government will be a conduit to a competitor's drawing board.

Current data license right regulations, and the resultant loss of exclusive control of data, also discourage the commercialization of military technology. A successful R&D program is only the first step in getting a product to market. Successful marketing and sales can be even more difficult.[40] A company has little incentive to take the risk and make the investment needed to establish, or even explore, a market niche if, when successful, a rival already has the same technology via the DoD and is ready to compete.

Small subtier firms argue that they are harmed more by data-rights regulations than are large prime contractors. Small companies often survive on one or a few products. Sometimes their only commercial advantage is a unique expertise in one particular technology, which, if compromised, could mean the end of the firm. The large primes have additional special "products" that they can sell: systems integration and the ability to deal with the government. Neither kind of information is as easy to steal as is a manufacturing process. Thus, the primes have relatively less worry about license rights. Moreover, large primes often require that data license rights clauses of their contracts are passed down to subcontractors; thus small subcontractors often view big prime contractors as part of the problem, not fellow sufferers. Small firms charge that the government is worse than cavalier about protecting data rights, that indeed the government sees any exclusive control of technology as a challenge to be met. Unless there are changes in requirements, many small firms will continue to be reluctant to make their technology available to the government.

Data license-rights questions do not lend themselves easily to compromise. Government and industry agree on what the issues are, but see a clear conflict between their respective interests. The government will always want to negotiate for as much access as it can get, and industry will always want to give up as little as possible. The optimal solution will require a broader perspective including the long-term effects on industry's incentives to provide the DoD with the products of its best technology and the DoD's long-term need to maintain some technologies regardless of short-term fluctuations in need or rate of production. (See ch. 4 for further discussion of technical data rights.)

Import and Export Restrictions

Import and export restrictions inhibit the entry of some companies into defense R&D in many indirect ways. Interviews show that some commercial firms are hesitant to take defense R&D contracts if the resulting technology is not readily exportable. The United States, unlike most other countries, often exports military technology to allies with the provision that further export to third countries will be controlled. This reduces incentives for international R&D collaboration among multinational and foreign firms.

Import restrictions affect U.S.-based firms that are truly multinational (as opposed to domestic firms with strong exports). Thus, a multinational corporation will balk at a government-sponsored development project—or require higher prices—if the resulting product must be manufactured in North America. A company like IBM makes products all over the world for a variety of economic reasons, and import restrictions (that is, Buy American rules)

[40] For many industries, getting a technically successful product accepted by the market is harder than the technical research and development itself. See Edwin Mansfield, "How Economists See R&D," *Harvard Business Review*, vol. 59, November-December 1981, pp. 100-106.

would require a complete overhaul in the way it does business.

If the United States wishes to increase civil and military technical integration, then it must reexamine its approach to the buying and selling abroad of technology for military applications. This is different from an arms export policy. Rather, the technical marketplace is becoming increasingly international; thus greater use of commercial technology inevitably leads to greater international interdependence. Indeed, as market barriers are reduced, tracking the origin of particular parts and technologies becomes increasingly difficult.

SUMMARY

Without offsetting action by the Federal Government or Congress, the DoD R&D budget will shrink in the future as the overall DoD budget declines. The DoD RDT&E is expected to drop in real terms from $40 billion today to $27 billion by 2001.

Yet the new international security environment still requires that the Nation have what *Redesigning Defense* termed an "advanced" military R&D capability that can respond to warnings of even sophisticated threats by supporting weapons systems that can meet the threat and be manufactured and deployed in time.

OTA defines an "advanced" DoD R&D capability as having

- a world-class manpower base (both individuals and teams);
- cutting-edge R&D able to guard against technological surprise, not only by sophisticated former adversaries, but by powers having access to the best weapons available on the international arms market;
- robust efforts in critical technologies;
- a balance between the near-term technology needs of each Service and the effort expended to meet the long-term R&D needs of U.S. defense overall;
- strong links to manufacturing, so the weapons systems proposed are producible and affordable; and
- strong links to civilian R&D, even in the absence of a national consensus about higher levels of Federal support for civilian technology programs.

Without offsetting actions, the likely shrinkage of DoD R&D will produce disproportionate cuts in private industry activities. Direct military-sponsored R&D in private companies will decline, as well as the investment private defense contractors make with their own funds.

Correspondingly, the fraction of military R&D effort done by Service in-house laboratories and FFRDCs would increase. While these institutions have a record of assisting the Services' direct needs they would have to change to address either research needs or the technology development role currently well performed by private companies.

Current and proposed plans to consolidate the Service's structure of laboratories and centers, while important, will not create the integrated management structure which the R&D component of the future DTIB will require.

The DoD may have to make a special effort to fully fund development work performed by private contractors to assure that technology development goes forward in private industry on profitable terms, even when there is unlikely to be a future production contract that would allow such companies to recover R&D costs. This would include support for prototyping, as discussed in chapter 3.

Without offsetting actions, performers of military R&D will not improve their links to civilian R&D. Present IR&D rules create barriers within companies between their military and civilian R&D efforts. Current technical data license-rights rules discourage specialized subtier firms—which are a critical source of the Nation's inventiveness in defense technology—from pursuing new technologies for both civilian and military use. Import and export restrictions inhibit interchange between defense and nondefense sectors and prevent the DoD from drawing on technology developed abroad, even by U.S.-based multinational firms.

Without offsetting actions, DoD support for research in colleges and universities could decline as the overall defense budget shrinks. Thus, the DoD will miss some of the benefits of basic university research it has enjoyed for many years. The DoD would also have less chance to train the next generation of scientists and engineers and familiarize them with the Nation's defense needs.

Chapter 3

Prototyping-Plus

Contents

Boxes

Figures

Tables

Chapter 3
Prototyping-Plus

INTRODUCTION

A challenge facing the Nation in the aftermath of the cold war is to reduce the size of the U.S. defense technology and industrial base (DTIB) while preserving key defense-related design and manufacturing teams, maintaining technological innovation, and giving the armed forces options from which to make future weapon-system and force-structure decisions. **One approach to this problem, called "prototyping-plus," would involve the continuous development of prototypes and, in selected cases, limited production for operational and field testing.** In the event of a need to replace obsolete systems or the emergence of a new military requirement, some of the prototyped systems could be further developed for quantity production.

Prototyping refers to the development and testing of working models—from computer simulations through operational hardware—to explore concepts and demonstrate specific design and operational objectives, thereby reducing technological uncertainties and risks. (See box 3-A.) The current weapons-acquisition process is based on the assumption that prototype development will lead in most cases to a design produced in quantity for the operational inventory. This assumption severely constrains the number of technological options that can be explored. A prototyping strategy, in contrast, would involve the exploration of a variety of system, subsystem, and component options *without* the assumption of proceeding to quantity production.

Greater reliance on prototyping at the expense of quantity production, as recommended by the Department of Defense (DoD), would have both benefits and costs. It would advance systems technology (e.g., systems design, not laboratory R&D), keep design teams intact, and support deployment of the most advanced equipment—assuming planners can see far enough into the future to begin production in a timely way. But it would sacrifice active forces and hot production lines, including large manufacturing teams. It is therefore necessary to define a new strategy that overcomes these drawbacks.

Defenders of the status quo often overlook the fact that the current acquisition system neglects the development of new manufacturing technologies, and that without a fundamental restructuring of the process, reduced procurement will further erode the DTIB. **The prototyping-plus approach would avoid simply putting new technologies "on the shelf" and allowing the manufacturing base to atrophy.**[1] Instead, design teams would hone their skills and know-how by developing and testing a series of prototypes, some of which could then be manufactured in limited quantities for field testing. By working out the major bugs in the manufacturing process, limited production would make it easier to negotiate the transition to quantity production—if and when such a decision is made. This approach could mitigate the effects of reduced procurement by replacing the boom-and-bust development cycle of the cold-war era with a more deliberate process, structured to preserve the full range of critical design, manufacturing, and support skills.

This chapter examines the feasibility of a prototyping-plus strategy and suggests how it might be implemented. The discussion addresses some frequent criticisms of this approach, such as the difficulty of moving from prototyping to quantity production; the unprofitability of prototyping; the problem of maintaining an adequate vendor base in the absence of significant production; the cost of prototyping; and the ability of a prototyping-plus strategy to preserve critical elements of the DTIB and its effects on jobs, skills, and training. The chapter also describes the larger restructuring of the DTIB that would be necessary for a prototyping-plus strategy to serve the Nation's future defense needs and to be profitable to all tiers of defense contractors.

THE PROTOTYPING SPECTRUM

Prototypes are useful in different ways depending on their role in the weapons-development process.[2] Figure 3-1 shows the different categories of proto-

[1] David Silverberg, "Acquisition Rule Irks Industry," *Defense News*, vol. 7, No. 6, Feb. 10, 1992, p. 10.

[2] I.C. Oelrich, Donald D. Weidhuner, and Frederick R. Riddell, *Small Turbine Technology Review*, IDA Paper P-1840 (Alexandria, VA: Institute for Defense Analyses, July 1985), p. 13.

Box 3-A—Traditional Functions of Prototyping

Prototyping has long served a number of functions in the weapons-acquisition process.

Hardware prototypes can define and reduce technological uncertainty in the development of a new system.[1] If the technological risks of a design are large and cannot be reduced by alternative techniques such as computer modeling or scale-model testing, construction of a working prototype is necessary. For example, vertical-takeoff-and-landing (VTOL) aircraft have complex aerodynamic and propulsion characteristics that are difficult to predict with analysis alone, so a prototype is needed to test performance predictions.

Prototypes can identify design flaws before a system enters full-scale development, also known as engineering and manufacturing development (EMD). A prototype nearly always reveals functional flaws in a design so that corrective action can be taken early. It is therefore possible to avoid the high costs and delays caused by engineering changes late in the development process or after production has begun.[2] During testing of the YA-10 prototype in 1974, for example, Fairchild discovered that during maneuvers at high angles of attack, the flow of air through the engine inlets was disturbed by turbulance from the fuselage-wing root area, causing the engines to flame out. The contractor used the prototype to develop and test a correction. In the absence of a prototype, this defect might not have been detected until the first production aircraft flew, when it could have caused a major crisis.[3]

Prototyping tests systems integration and exposes problems with electromagnetic interference and compatibility (EMI/EMC) and software. General Dynamics first bench-tested the M1A2 tank's digital mapping system and other electronic subsystems individually. They were then integrated in a laboratory, tested in a technology demonstrator, and finally put in a prototype tank. Even so, it took months of testing to correct operational discrepancies and to debug the software.[4] It is not enough to test various subsystems in the laboratory; in many cases, they must be integrated in a prototype and tested under realistic conditions.

Prototyping can help define how to accomplish a given military mission before a production decision is made. Prototypes can test out different approaches to performing a given mission (e.g., ballistic-missile defense). The experience gained in prototyping can then lead to faster and lower cost completion of development and production.

Competitive prototyping can help to select a prime contractor. Competitive prototyping led the Army to select a different contractor for the AH-64 attack helicopter than it would have chosen based on the original paper proposals. During the paper competition, many program personnel believed that Bell Helicopter had a better design. But in the prototyping phase the Hughes Aviation prototype outperformed Bell's, and conceptual differences between the two designs were resolved in Hughes' favor. As a result, the Army awarded Hughes the contract.[5]

Prototyping tests the soldier/system interface for the first time. The man in the loop remains the most essential ingredient of successful hardware/software development. In some cases, problems in the soldier/system interface cannot be identified and corrected early without prototyping.

---

[1] B.H. Klein, T.K. Glennan, Jr., and G.H. Shubert, *The Role of Prototypes in Development*, RM-3467/1-PR (Santa Monica, CA: RAND Corp., April 1971), p. 10.

[2] Robert Perry, *A Prototype Strategy for Aircraft Development*, RM-5597-1-PR (Santa Monica, CA: RAND Corp., July 1972), p. 9.

[3] G.K. Smith et al., *The Use of Prototypes in Weapon System Development*, R-2345-AF (Santa Monica, CA: RAND Project Air Force, March 1981), p. 58.

[4] Interview with William F. Cody, corporate director of Land Systems, General Dynamics, Washington office, Nov. 13, 1991.

[5] Smith et al, op. cit., footnote 3, p. 166.

types, positioned along a spectrum from the concept-definition phase to the engineering and manufacturing development (EMD) phase. Each kind of prototype is discussed in detail in the following sections.

Conceptual Prototypes

Conceptual prototypes are computer simulations of hypothetical systems. During exploratory development, simulation can emulate the capabilities and properties of an aircraft or a tank that exists only in the computer's memory. (See box 3-B.) Simulators generate dynamic visual environments that are impressively realistic, enabling military users to practice aerial dogfights or tank engagements, complete with simulated terrain, smoke, and enemy vehicles. The Defense Advanced Research Projects Agency (DARPA) has developed a Simulation Network (SIMNET) consisting of 120 computer-controlled and networked simulators of M1A1 tanks, Bradley infantry vehicles, helicopters, and fighter-bombers located at military bases throughout

Figure 3-1—The Prototyping Spectrum

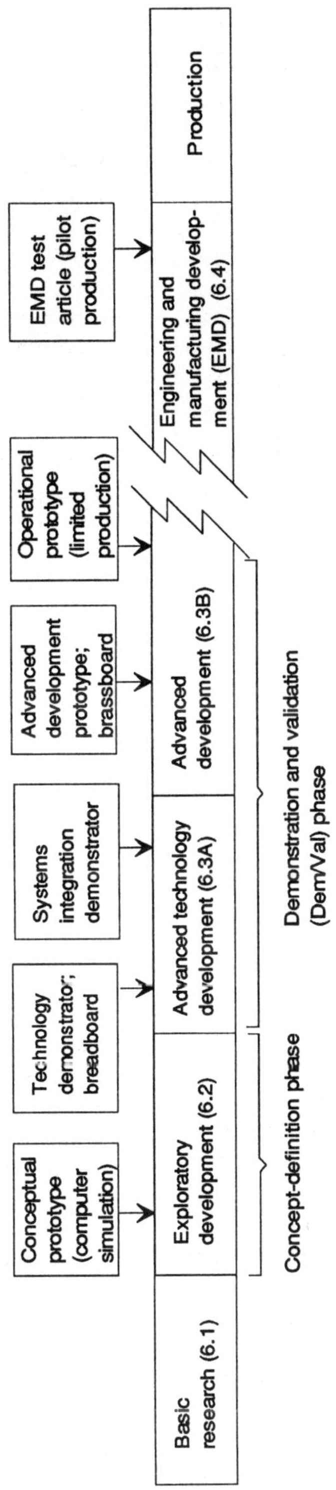

SOURCE: Office of Technology Assessment, 1992.

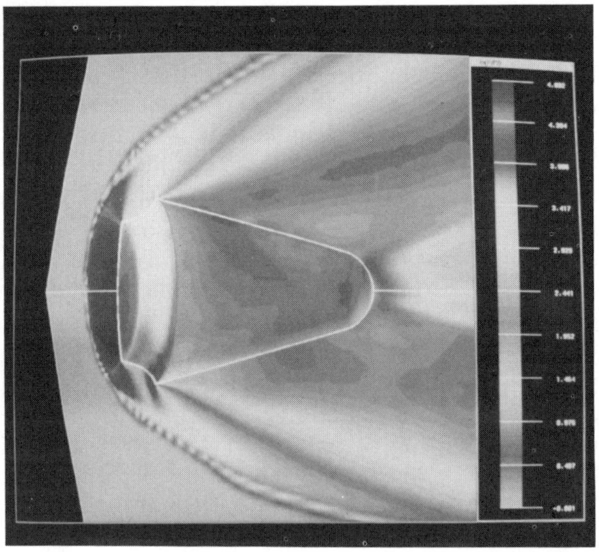

Photo credit: Rockwell International

Computational fluid dynamics (CFD) simulates the aerodynamics of reentry of a proposed single-stage-to-orbit rocket.

Photo credit: DoD

A Lockheed technician models various aircraft concepts on a computer-graphics generator.

the United States and Europe. These interlinked simulators can be used to fight imaginary war games.

Interactive computer simulations can also inform and focus the definition of new military systems in advance of hardware development by evaluating the effects on military performance of proposed design changes. Such models can help planners sort through various threat scenarios and assess which new technologies and capabilities would provide the greatest payoff on the battlefield. For example, DARPA has sponsored the development of an interactive simulation called Project Odin, which reconstructs a pivotal tank battle during the Gulf War between the U.S. Army and the Iraqi Republican Guard. The simulation is highly detailed, including the characteristics of the weapon systems on both sides, as well as sight lines, damage, and casualties.[3] Parameters of friendly and enemy weapon systems can be altered interactively to assess the impact on the outcome of the battle if, say, the Iraqi tanks had been equipped with thermal sights, or U.S. tank guns had had 20 percent more range. (The latter simulation might reveal, for example, that increasing the firing range of U.S. tank guns would offer no operational benefit unless they had improved thermal sights that could acquire targets at greater distances.)

Conceptual prototyping has limitations. Some types of aerodynamic behavior are so complex that a physical prototype must be tested before a design concept can be validated. Other tasks exceed the capabilities of computer simulation, such as integrating multiple subsystems into a platform or using new materials with unknown aging and fatigue characteristics. There are also "unknown unknowns" —phenomena whose existence is unsuspected until they emerge in testing. Further, interactive simulation often does not account for training, morale, or unexpected enemy tactics.[4]

Technology Demonstrators

A *technology demonstrator* is a functional vehicle (or test rig) that is built and tested to answer a few important technical questions as cheaply as possible. It can provide the proof-of-principle of an enabling technology or design configuration, or explore in a preliminary way the characteristics of a new systems

[3] F. Clifton Berry, Jr., "Re-creating History: The Battle of 73 Easting," *National Defense*, vol. 76, No. 472, pp. 6-9.

[4] After World War II, Admiral Nimitz commented on war planning: "We had war-gamed every single possibility of how and what the Japanese would do in the Pacific, and we were ready for it . . . . All except one: we never expected them to use the kamikaze tactic."

Box 3-B—Computer Simulation as an Analytical Tool

In earlier years, computers were used to speed analytical calculations during system design and to process data derived from empirical studies. Today, however, computers also have begun to replace drawing boards and wind tunnels for purposes of design and analysis. Most aerospace engineers use computer-aided design (CAD) for drafting, and an increasing number rely on computer-aided engineering (CAE) for structural and physical analysis.

CAE uses computational models to simulate the behavior of hypothetical systems. For example, finite element analysis models the stresses in a complex structure, like an aircraft wing, by representing the object as a collection of discrete elements with specified properties. Computational fluid dynamics (CFD) simulates the flow of air or water over a body (e.g., a plane or submarine). It can greatly reduce the time devoted to costly wind-tunnel testing. Finally, computer simulations can integrate "human-factors engineering" into the design, manufacture, operation, and maintenance of weapon systems to improve compatibility between people and machines.[1]

In a growing number of cases, computer simulation can dispense with the need for a complex test article to emulate real-world conditions. For example, CFD is more accurate than wind-tunnel testing for the simulation of unsteady flow conditions within a jet engine or for a jet fighter flying at high angles of attack. Computers can also simulate velocities and environments of hypersonic flight vehicles that cannot be duplicated by traditional wind-tunnel studies.[2] Yet computer-based simulation tools are far from perfect. While supercomputers can simulate the aerodynamic behavior of a hypothetical aircraft, the simulations are only as good as the computational model on which they are based. Further, computational complexity tends to increase costs as software models become more elaborate.

The limitations of computer simulation often make hardware prototyping necessary. Such prototypes have advantages in testing an overall system and identifying manufacturing problems. They can also allow engineers to verify computational models like CFD by correlating them with real-world physical phenomena. Moreover, before engineers simulate an entirely new phenomenon such as stealth, a prototype can help build a database on how radar and detection technologies are affected by different shapes, textures, aspect angles, and electromagnetic properties.

One strategy for reducing total development costs in the future smaller DTIB would be to combine computer simulation with limited hardware prototyping. Cost constraints already require the use of computer simulation during tactical-missile development. A software program first simulates the engagement between the missile and target, and tests the performance of the missile's seeker and guidance computer. Then a limited number of hardware prototypes are fired against a set of targets selected to verify the computer model.[3] Once verified, the model can be used with confidence to explore system performance throughout the engagement envelope. This approach could be applied to other systems in development, quite apart from any decision on production.

[1] William B. Scott, "Computer Simulations Place Models of Humans in Realistic Scenarios," *Aviation Week & Space Technology*, vol. 134, No. 25, June 24, 1991, pp. 64-65.

[2] Dean R. Chapman, "A Perspective on Aerospace CFD," *Aerospace America*, vol. 30, No. 1, January 1992, pp. 19, 58.

[3] Telephone interview with Donald Putnam, corporate director of Contracts and Technical Analysis, General Dynamics, Jan. 22, 1992.

concept.[5] Technology demonstrators are also built for subsystems, such as the thrust-vectoring engine nozzle developed by Pratt & Whitney. A technology demonstrator of an electronic subsystem, built and tested in a laboratory, is known as a *breadboard.*

The best-known technology demonstrators are the series of experimental "X" vehicles, built intermittently by U.S. aerospace companies since the late 1940s for the Air Force, NASA, or DARPA. (See table 3-1.) An X-plane is often little more than an airframe, engines, and flight controls, without the specialized electronics and integrated armaments required for an operational weapon system. The X-31 demonstrator, for example, was developed to explore new technologies to enhance fighter maneuverability and does not include many subsystems required for a combat-capable aircraft. Technology demonstrators incorporate a few custom-built elements essential to the concept or technology being

[5] According to the DoD, advanced technology transition demonstrators (ATTDs) are intended to test "integrated technologies in as realistic an operational environment as possible to assess the performance payoff or cost-reduction potential of advanced technology before program-specific prototyping begins." Under Secretary of Defense for Acquisition, *Department of Defense Directive No. 5000.1*, Feb. 23, 1991, p. 5-C-2.

Table 3-1—The X-Aircraft and Missiles, 1946-1991

X-plane	Company	1st flight	Mission
X-1	Bell	01/25/46	Identify dynamic flight characteristics of supersonic aircraft.
X-1A	Bell	02/14/53	Investigate aerodynamic phenomena at speeds greater than Mach 2 and altitudes above 90,000 feet.
X-1E	Bell	12/12/55	Explore potential performance improvements to Mach 2.5.
X-2	Bell	06/27/52	Build swept-wing version of X-1 to achieve higher speeds and altitudes, investigate aerodynamic heating.
X-3 *Stiletto*	Douglas	10/20/52	Explore high-speed flight with takeoff and landing under own power, and low-aspect-ratio wings.
X-4 *Bantam*	Northrop	08/18/50	Test aircraft design without horizontal tail at trans-sonic speeds.
X-5	Bell	06/20/51	Investigate aerodynamics of variable-sweep-wing aircraft.
X-6	Convair	canceled 1953	Investigate operational feasibility of nuclear propulsion systems prior to commitment to prototype military nuclear-powered aircraft.
X-7A, B	Lockheed	04/26/51	Build testbed for supersonic and hypersonic ram jet engine.
X-8 *Aerobee*	Aerojet	11/24/47	Develop inexpensive upper-atmospheric research vehicle/sounding rocket with parachute recovery system.
X-9 *Shrike*	Bell	04/28/49	Build simplified testbed for air-to-surface missile to obtain data on aerodynamics, stability, propulsion, and servo and guidance systems.
X-10	North American	10/14/53	Build aerodynamic and systems testbed for the *Navaho* cruise-missile program.
X-11	Convair	06/11/57	Develop single-stage ballistic rocket to obtain design data for the planned *Atlas* intercontinental ballistic missile.
X-12	Convair	07/09/58	Build high-performance one-and-a-half stage ballistic missile to prove systems and hardware configuration for production version of the *Atlas* missile.
X-13 *Vertijet*	Ryan	12/10/55	Explore feasibility of building a pure jet vertical-takeoff-and-landing (VTOL) fighter.
X-14/A, B	Bell	02/17/57	Study experience of a pilot flying a VTOL aircraft from a normal crew station using standard aircraft flight references.
X-15/X-15A-2	North American	06/08/59	Investigate problems of atmospheric and space flight at very high speeds and altitudes (Mach 6.6 and 250,000 feet).
X-16	Bell	canceled 1955	Build high-altitude, long-range reconnaissance aircraft carrying various sensors. (Replaced by Lockheed U-2.)
X-17	Lockheed	04/17/56	Build multistage rocket to transport various reentry-vehicle configurations to very high altitudes for testing.
X-18	Hiller	11/24/59	Assess feasibility and practicality of large, tilt-wing VTOL aircraft.
X-19	Curtiss-Wright	11/20/63	Demonstrate tilt-propellor VTOL configuration for transition from hover to forward flight.
X-20 *Dyna-Soar*	Boeing	canceled 12/10/63	Provide a manned, maneuverable vehicle to collect data on controlled reentry from orbital flight.
X-21A	Northrop	04/18/63	Explore feasibility of full-scale boundary-layer control on large, subsonic aircraft.
X-22A	Bell Aerospace Textron	03/17/66	Evaluate dual-tandem ducted propellor configuration for V/STOL aircraft.
X-23A *Prime*	Martin Marietta	12/21/66	Test configurations, control systems, and ablative materials for hypersonic lifting-body type reentry vehicles.

(continued on next page)

Table 3-1—The X-Aircraft and Missiles, 1946-1991—Continued

X-plane	Company	1st flight	Mission
X-24/A-C	Martin Marietta	04/17/69	Explore low-speed flight characteristics of maneuverable lifting-body design.
X-25/A,B	Bensen	01/23/68	Build small, ultralight aircraft to provide emergency egress capabilities beyond those of a conventional parachute.
X-26B	Lockheed	07/67	Develop quiet plane to carry dedicated sensors over enemy territory to obtain real-time intelligence during the Vietnam War.
X-27 *Lancer*	Lockheed	canceled 1971	Build prototype of advanced, lightweight fighter to replace F-104, with potential for both U.S. and foreign sales.
X-28A *Osprey*	Pereira	08/12/70	Explore potential usefulness of a small, single-seat seaplane for civil police patrol duty in Southeast Asia.
X-29A	Grumman	12/14/84	Assess benefits and costs of forward-swept wing, relaxed static stability, and related technologies.
X-30A *NASP*	Rockwell	1999 (est.)	Build hardware testbed for National Aerospace Plane (NASP) with single-stage-to-orbit capability.
X-31A	Rockwell/MBB	10/11/90	Break the so-called "stall barrier" to permit close-in aerial combat beyond normal stall angles-of-attack.

SOURCE: Jay Miller, *The X-Planes: X-1 to X-31* (Arlington, TX: Aerofax, 1988).

demonstrated, but make extensive use of off-the-shelf hardware. Thus, more than 50 percent of the X-31 consists of government-furnished equipment from other aircraft.[6]

The other Services have also built technology demonstrators. The Army's Advanced Composite Airframe Program demonstrated that primary aircraft structures could be made of composite materials and led to the use of composites in the V-22 Osprey aircraft. In the mid-1980s, General Dynamics Land Systems Division developed the Tank Test Bed, an experimental armored vehicle that featured an unmanned gun turret operated by remote control. Currently, General Dynamics is developing the Composite Armored Vehicle, which will explore radically new armors and manufacturing methods.[7] Navy technology demonstrators have included a quiet torpedo-launching system and a stealthy warship design to reduce vulnerability to enemy radars and guided missiles.[8]

There are two other kinds of technology demonstrators. A *technology integration demonstrator* assembles available, off-the-shelf subsystems to perform a unique mission. For example, the Advanced Fighter Technology Integration (AFTI) program in 1983-84 modified an F-16 to demonstrate technologies that could improve fighter maneuverability. A *production retrofit demonstrator* is an upgrade of an existing platform that incorporates some new capability. For example, earlier models of the F-15 were used to test new subsystems that were incorporated into the F-15E.

The history of the X-aircraft shows that technology demonstrators can provide a leg up on next-generation systems, often in a serendipitous manner. Table 3-2 indicates technologies from six X-aircraft programs that found their way into weapon systems, although many of the design concepts were so revolutionary that they were not applied for decades. Similarly, Northrop developed a number of "flying wing" technology demonstrators in the late 1940s.[9] Although the flying-wing program was later cancelled, flight-testing of the prototypes gave Northrop an extensive database on the aerodynamic coefficients, stability, and range/payload characteristics of these exotic designs. When Northrop

[6] Off-the-shelf subsystems in the X-31 include the General Electric F404 engine, the canopy and windscreen from the F-18, the landing gear from the F-16, the wheels and brakes from the Cessna Citation III, and derivatives of existing Honeywell computers. Brian Wanstall and J.R. Wilson, "Air Combat Beyond the Stall," *Interavia Aerospace Review*, No. 5, June 1990, p. 406.

[7] Telephone interview with Otto Renius, chief scientist, General Dynamics Land Systems Division, Sterling Heights, Michigan, Dec. 10, 1991.

[8] Robert Holzer and Neil Munro, "Navy Invests Over $1 Billion in Stealth Ship," *Defense News*, vol. 7, No. 4, Jan. 27, 1992, p. 1.

[9] A propellor-driven version called the XB-35 was first flown in June 1946, and a jet-powered version called the YB-49 was first flown in October 1947. Christopher Chant, *Aircraft Prototypes* (Seacaucus, NJ: Chartwell Books, 1990), p. 8.

Photo credit: Pratt & Whitney

Technological evolution of a thrust-vectoring jet engine nozzle. A "boiler-plate" nozzle *(left)* provided basic mechanical and thermal data, which were incorporated into a durability demonstrator *(top center)*. Initial flight testing was performed with a technology demonstrator *(right)*. Finally, lessons learned in manufacturing and flight testing were applied in an advanced-development prototype *(bottom center)*.

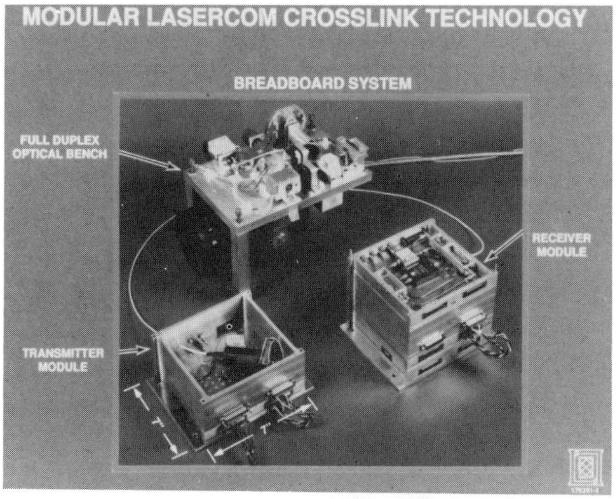

Photo credit: Lincoln Laboratories

"Breadboard" version of a modular laser-communications systems was built for lab testing. It can be developed further into an advanced-development prototype, or "brassboard," for operational testing.

Table 3-2—Technological Spinoffs of X-Aircraft Programs

X-aircraft	1st flight	Program goal	Beneficiary program/date
X-1	01/25/46	Supersonic flight	F-100 (1953)
X-4	08/18/50	Tailless aircraft	F-102 (1953) F-106 (1956)
X-5	06/20/51	Variable-sweep wings	F-111 (1964) F-14 (1970) B-1 (1974)
X-15	06/08/59	Hypersonic flight and spaceflight	SR-71 (c. 1964) Space Shuttle (1981)
X-23/24	12/21/66	Hypersonic lifting-body concept and materials	Space Shuttle (1981)
X-29	12/14/84	Relaxed stability, composite wings, forward-swept wings	ATF (mid-1990s)

SOURCE: Rockwell International, "X-Planes: Aeronautical Research Tools Have Paid Big Dividends in U.S. Aviation Leadership: A Perspective."

decided in 1979 to use a flying-wing configuration for the B-2 strategic bomber because of its superior stealth characteristics, the company turned to the technical database collected some 30 years earlier.[10]

A technology demonstrator sometimes achieves a major breakthrough in performance that spurs a procurement decision that would not otherwise have been made. Historical cases include the U-2 and the SR-71 reconnaissance aircraft, developed secretly by the Lockheed Skunk Works for the Central Intelligence Agency. A more recent example is the Lockheed HAVE BLUE stealth-technology demonstrator, sponsored by DARPA. This $43 million program demonstrated the use of a faceted airframe design to minimize radar reflections. (The design was faceted rather than curved because of limits on computing power at the time.) The HAVE BLUE program built two small, nonmissionized technology demonstrators that weighed only 12,000 pounds fully loaded and used many components from existing aircraft. The first of the two prototype

[10] Telephone interview with George J. Friedman, vice president for Engineering and Long-Range Planning, Northrop Corp., Dec. 16, 1991.

Photo credit: Northrop Corp.

"Flying wing" technology demonstrator, the YB-49 *(left)*, was first flown by Northrop in 1947. Three decades later, the company applied flight-test data from the YB-49 to develop another flying wing, the B-2 bomber *(right)*.

aircraft flew in early 1978, after 20 months of development; both were flight-tested for 18 months.[11] Lockheed demonstrated that the faceted configuration could fly and that the aircraft's radar signature was as low as predicted, although both HAVE BLUE aircraft crashed during flight testing.[12] In December 1978, the Air Force moved the program directly into the engineering and manufacturing development (EMD) phase. Lockheed then implemented the stealth technologies developed for HAVE BLUE in an operational fighter-attack aircraft, the F-117.

Although the initial 28 aircraft in the X-series had their first flights between 1946 and 1970, there was a hiatus of 14 years, from 1970 to 1984, between the X-28 and the X-29. (The HAVE BLUE was not officially an X-aircraft, although it met the same criteria.) In 1986, the President's Blue Ribbon Commission on Defense Management (the Packard Commission) expressed concern about the drop in the number of demonstrator programs. The Commission recommended "a high priority on building and testing prototype systems to demonstrate that new technology can substantially improve military capability, and to provide a basis for realistic cost estimates prior to a full-scale development decision."[13]

Since the early 1980s, there has been a modest resurgence of interest in experimental aircraft. The Grumman X-29 demonstrator, which first flew in December 1984, sought to enhance fighter maneuverability by integrating forward-swept wings, canards, composite structures, and flight-control software for inherently unstable aircraft. The Rockwell-MBB X-31, which first flew in October 1990, also tried to improve fighter maneuverability through the use of integrated control systems and a thrust-vectoring engine.

Since technology demonstrators are designed primarily to provide information, they are of most value if they give clear positive or negative answers

[11] Bill Sweetman, "Lifting the Curtain: Stealth Techniques Detailed," *International Defense Review*, vol. 25, No. 2, February 1992, p. 159.

[12] While neither crash was the result of the low- observable technology, a hazardous design flaw was detected and removed. Jack S. Gordon, Lockheed Advanced Development Co., personal communication.

[13] The President's Blue Ribbon Commission on Defense Management, *A Quest for Excellence: Final Report to the President*, June 1986. Reprinted in U.S. Congress, House Committee on Armed Services, *Defense Acquisition: Major U.S. Commission Reports (1949-1988), Volume I* (Washington, DC: U.S. Government Printing Office, Nov. 1, 1988), p. 937.

326-447 - 92 - 3 : QL 3

Photo credit: DoD

X-29 technology demonstrator explored the use of forward-swept wings, canards, and an inherently unstable configuration to enhance fighter maneuverability.

to functional, operational, or manufacturing questions. Nevertheless, many useful defense technologies were not developed for specific applications or were applied in ways that the original inventors did not imagine. A good example is laser-based guidance, which was initially developed by the U.S. Army Missile Command for antitank missiles. The Army became disenchanted with the technology and transferred it to the Air Force, which applied it to the development of the laser-guided bomb in the 1960s.[14] Future demonstrator programs might therefore seek a balance between "technology push," or the pursuit of technological innovation for its own sake, and "technology pull," or more focused development efforts disciplined by a clear mission application and schedule requirement.

Box 3-C—Submarine Prototypes

The 1950s saw rapid innovation in submarine design and construction. Submarines changed from being primarily surface boats that submerged occasionally to being capable of nearly unlimited endurance under water. Prototyping played a major role in this evolution. The USS *Albacore* (commissioned in 1953) was a technology demonstrator that tested a streamlined hull shape and novel steering devices. The first two nuclear-powered submarines were advanced-development prototypes built for operational deployment. The *Nautilus* (commissioned in 1954) had a reactor cooled with water, whereas the original *Seawolf* (commissioned in 1957) had a reactor cooled with liquid sodium. The water-cooled reactor was eventually judged superior; all U.S. naval reactors since then have been water-cooled.

Technological innovations were integrated into 7 different submarine prototypes built between 1956 and 1960, all of which entered the operational fleet. Most U.S. submarines, however, were produced in multiple copies, including 4 *Skate* class, 6 *Skipjack* class, 14 *Thresher/Permit* class, 37 *Sturgeon* class, and 55 *Los Angeles* class. Since the new SSN 21 *Seawolf* will be canceled after production of only one, or possibly two or three units, this submarine will effectively be a prototype. It will join the operational fleet and serve as an R&D test bed. The Navy's proposed *Centurion*, a candidate next-generation attack submarine, is envisioned as a low-cost, modular system.[1]

[1] Barbara Starr, "Lone Seawolf to join USN fleet," *Jane's Defence Weekly*, Feb. 29, 1992, p. 339.

Advanced-Development Prototypes

During the demonstration and validation (dem/val) phase, *advanced-development prototypes* are often built to determine whether the chosen configuration can meet program objectives in terms of performance, cost, or operational suitability. Even negative answers are useful, since they can help to avoid technological dead-ends.[15] Advanced-development prototypes of electronic subsystems, called *brassboards*, are designed to be tested in an operational environment. Large weapon systems have sometimes been prototyped as single units, which were later deployed as operational combatants. During the 1950s, for example, the U.S. Navy developed several one-of-a-kind prototypes of submarines (box 3-C), as well as nuclear-powered cruisers and aircraft carriers.

Whereas a technology demonstrator seeks to answer a basic technical question, an advanced-development prototype is the first physical representation of a potential operational system. There are two reasons for building an advanced-development prototype: to demonstrate through testing that the product has the required capabilities, and to estimate the time and cost of producing the system, along

[14] Peter deLeon, *The Laser-Guided Bomb: Case History of a Development*, R-1312-1-PR (Santa Monica, CA: RAND Corp., June 1974), pp. 6-10.

[15] Karen W. Tyson, et al., *Acquiring Major Systems: Cost and Schedule Trends and Acquisition Initiative Effectiveness*, IDA Paper P-2201 (Alexandria, VA: Institute for Defense Analyses, March 1989), p. VIII-1.

with its manufacturability and maintainability.[16] The validation of manufacturing processes and cost may require extrapolating from the prototyping experience into the actual production environment, using factory personnel and equipment. A novel approach to this problem is to develop computer simulations of manufacturing.

In sum, the information generated by a software or hardware prototype depends on its role in the development process. A computer simulation or technology demonstrator usually evaluates some limited design parameters, whereas an advanced-development prototype offers greater fidelity to the final production system but costs much more. The closer a prototype corresponds to the production model, the more it is locked into assumptions about the nature of military threats—assumptions that may be called into question in the future. Thus, the choice of which class of prototype to build is determined by such factors as the maturity of product and manufacturing-process technologies, the degree of uncertainty in the security environment, and the need to preserve technical competence and to maintain production capacity.

ASPECTS OF A PROTOTYPING-PLUS STRATEGY

A prototyping-plus strategy would involve the following elements, as illustrated in figure 3-2.

Increased development of prototypes. Prototyping would maintain the U.S. edge in defense technology for major systems (ships, aircraft, tanks, etc.) despite cuts in production and new program starts. Analyses of emerging military threats and computer simulations would identify new capabilities that might provide a clear performance advantage at an acceptable cost. A technology-demonstrator program could then be launched without a formal military requirement or the assumption of an eventual procurement.

Building a technology demonstrator might involve only one design team, or might involve competition between two or more industrial teams. In competitive prototyping, at least two technologically distinct systems would be built for testing, and one would then be chosen for further development or

Photo credit: DoD

Two advanced-development prototypes of the F-22 Advanced Tactical Fighter during flight testing.

production. Competition in selected areas might make each firm or industrial team more productive and hence improve quality and contain costs; competitive prototyping that considers dissimilar designs might also hedge against new technologies and threats. Nevertheless, funding constraints may restrict the use of competition to relatively inexpensive demonstrators rather than advanced-development prototypes.

Production of operational prototypes. Firms might manufacture a limited number of operational prototypes of one design to validate performance, manufacturing processes and controls, and projected costs. These systems would be designed for producibility and would include enough armaments, fire-control, and other subsystems to give them some operational capability. Military users would then put the prototypes through trials, since a new military capability cannot be realized until servicemen test it out under realistic field conditions.

Enough operational prototypes would be produced to enable military customers to

1. develop tactics and doctrines;
2. perform reliability, maintenance, and live-fire testing; and
3. provide feedback to the development team on improvements needed to fine-tune the system and compensate for operational shortcomings.[17]

[16] Defense Systems Management College, *Department of Defense Manufacturing Management Handbook for Program Managers*, 2d ed. (Fort Belvoir, VA: Defense Systems Management College, July 1984), p. 3-5.

[17] Gordon R. England, "Statement Before the House Armed Services Committee Structure of U.S. Defense Industrial Base Panel", Oklahoma City Field Hearing, Nov. 1, 1991, pp. 7-8.

Figure 3-2—Prototyping-Plus Strategy

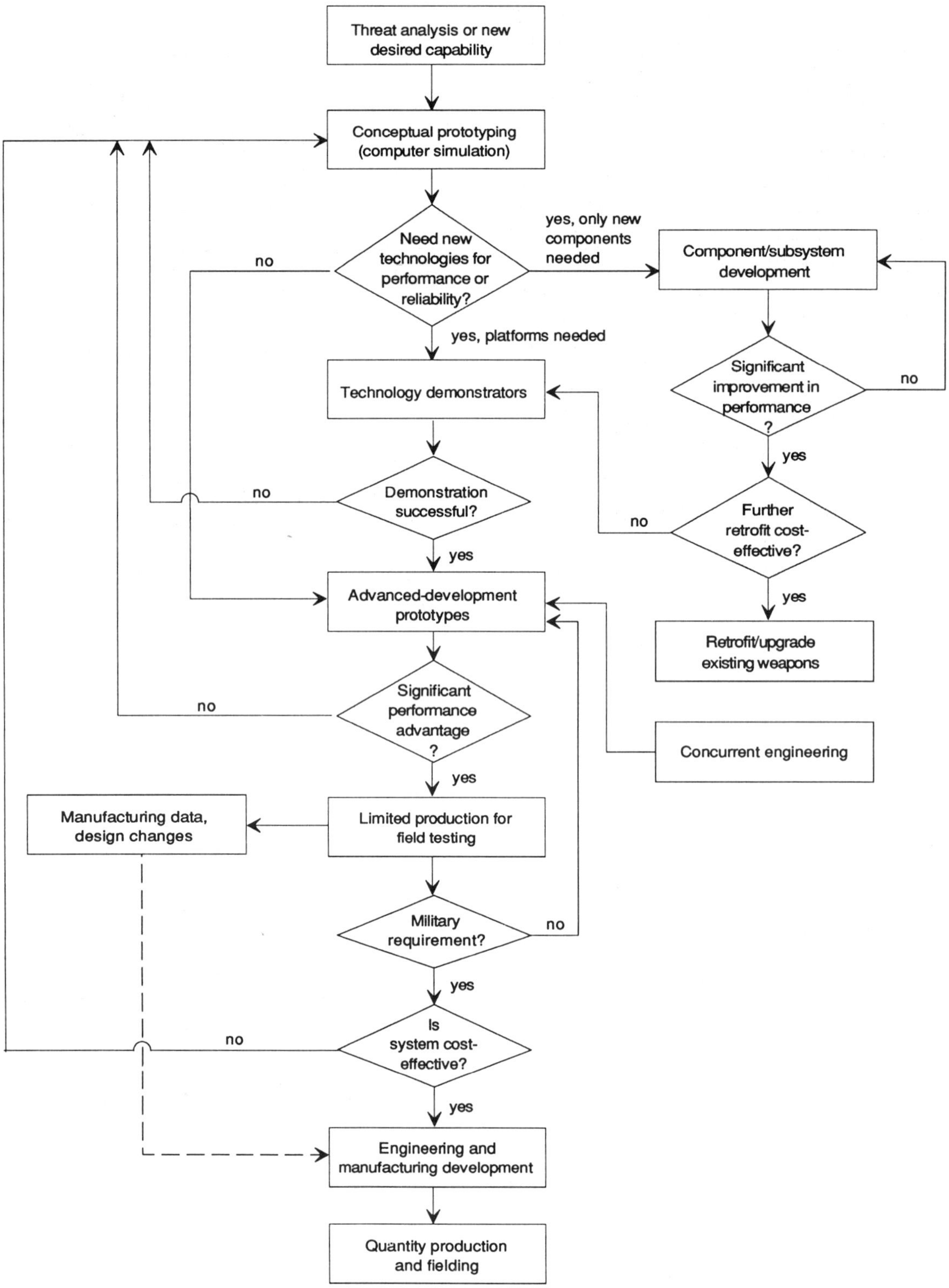

SOURCE: Office of Technology Assessment, 1992.

For these purposes, it might be sufficient to build a platoon of tanks or a squadron of aircraft.[18]

Limited production of prototypes would also provide some preliminary manufacturing data, increasing industry's ability to produce the system when needed, in sufficient quality, and at a target cost. Since long production runs would not be available to improve poor designs, a prototyping-plus strategy would emphasize designing for producibility, moving forward production issues that currently are not addressed until much later in the development process. Thus, **a prototyping-plus strategy would achieve a marriage of R&D and manufacturing, with the goal of supporting both.**

Limited production of prototypes raises the issue of how a small number of unique systems would be supported logistically in the field. In the past, the Services have provided logistical support for small numbers of complex systems, including the U-2 and SR-71 aircraft and various "testbed" vehicles fielded by the Army's Ninth Infantry Division. Logistical support could be contracted to the same firm that produced the prototype, rather than breaking out spare-parts production for competitive bid. This approach would minimize the impact of limited production on the DoD's logistical system. But it would require modifying the current procurement regulations mandating "free and open competition," as discussed in chapter 4.

Selective replacement of major systems. Prototypes would preserve the *potential* to move into quantity production when needed, although only a fraction of all prototypes would enter the engineering and manufacturing development (EMD) phase. Quantity production could be ordered when

1. a radically new technology is developed (e.g., stealth) that cannot be retrofitted into a current platform;
2. a new or emerging threat warrants a new deployment; or
3. the current system has aged to the point where replacement is more cost-effective than an upgrade.[19]

To go to full production, the Services would need to demonstrate a real requirement. The production contract could either be awarded to the same firm that designed the prototype, or opened up for competitive bid.

Systems v. Components

A prototyping-plus strategy could consist of two parallel but interlinked tracks, one focused on components and subsystems and the other aimed at new platforms. Although the discussion of prototyping has concentrated largely on platforms, it would be more cost-effective to emphasize the development of improved subsystems (such as cockpit displays, mission computers, night-vision sights, and airborne radars), which could be retrofitted at regular intervals into fielded platforms to achieve improvements in performance. Component or subsystem development could be accompanied by development and validation of the manufacturing processes needed to produce them.[20]

In considering a prototyping-plus strategy, the DoD should strive for an optimal balance between upgrading fielded weapons and developing new systems for the next century. There is a need to change the mentality in the R&D community to make product improvement the first priority. At the same time, the new platform prototypes could make maximum use of the improved components and subsystems being developed on a second track. For example, several new components and subsystems could be integrated into a new system prototype, setting the stage for force modernization if and when a requirement for the new item emerges. To this end, the Services might jointly develop modular subsystems to be inserted into different weapon systems. An example of this approach is the Joint Integrated Avionics Working Group (JIAWG), a tri-Service office created to develop common avionics modules for the Air Force's Advanced Tactical Fighter, the

[18] A U.S. Army platoon has 5 tanks. A U.S. Air Force wing typically consists of 3 squadrons, each containing between 18 and 24 aircraft (depending on type).

[19] For example, in the case of the F-15 fighter, one could argue that neither the age of the aircraft nor the threat warrants near-term replacement with a more modern fighter. Stealth technology might justify a wholesale replacement, but only if it were a critical factor in the execution of F-15 missions. Thus, the current absence of a significant threat and the reduced wear on the F-15 force in the post-cold-war era may provide a sufficient basis *not* to produce a follow-on weapon system for several years.

[20] John D. Morrocco, "Dangers Cited in Implementing New Pentagon Acquisition Strategy," *Aviation Week & Space Technology*, vol. 136, No. 10, March 9, 1992, p. 21.

Navy's A-12 strike aircraft, and the Army's Comanche helicopter.[21]

A prototyping-plus strategy would be compatible with either an evolutionary or revolutionary approach to weapons development. The lack of a large-scale military threat to U.S. security gives the Nation the freedom to emphasize either the acquisition of knowledge and technology for future advances in military performance, reliability, and maintainability, or the evolutionary upgrading of fielded systems. Thus, a first-generation prototype aircraft might focus on demonstrating incremental improvements in maneuverability or target acquisition, while the next-generation system could aim at entirely new capabilities such as stealth. When prototypes do not go into full production, the technology they embody could be recycled into other systems.

Profitability of Prototyping

A prototyping-plus strategy would require a significant change in attitude from both government and industry. When procurement budgets were large, companies were generally willing to break even or even lose money on R&D in the expectation of making profits on a follow-on production contract. As a result, the DoD could get private firms to provide a large share of the development funding. The result was to understate the true cost of design and development.

At present, defense firms are unable or unwilling to invest their own money and engineering resources in prototypes that may not enter quantity production for years, if ever. The case of the Army's proposed Mobile Protected Weapon System, a light tank to be deployed by parachute from a transport aircraft, indicates why. In 1980, the Army announced it would buy 300 of these tanks and invited industry to propose systems that met its specifications. Three U.S. producers of armored combat vehicles—FMC Corp., Teledyne's Continental Motors Division, and Cadillac Gage—responded by each building prototypes at their own expense, at a cost of $20 to 25 million per prototype. Subsequently, the Army cancelled the program.[22] Although Cadillac Gage sold a modified version of its prototype to Thailand, the other two firms had to write off their investments.[23]

Since private-sector firms lack economic incentives to finance prototypes on their own, the government will have to bear most if not all the costs of prototyping. Prototyping-plus would be compatible with a U.S. defense industry made up of fewer companies. These firms would have to downsize significantly while maintaining their core R&D and manufacturing capabilities, including design teams. Nevertheless, prototyping-plus would not be sufficient to preserve the defense production base. Since prototyping involves relatively little manufacturing, other measures would have to be taken to preserve manufacturing know-how. Moreover, manufacturing firms cannot be expected to survive entirely on prototyping contracts alone. **A prototyping-plus strategy would only be viable in conjunction with an integrated restructuring of the DTIB, including low-rate production, retrofits, and greater integration with the civil sector.**

IMPLEMENTING THE STRATEGY

A prototyping-plus strategy should

1. keep design teams intact and technologically competitive by continually updating their skills;
2. help preserve essential manufacturing know-how;
3. facilitate the transition from prototyping to quantity production when a procurement decision is made, given sufficient lead-time and adequate funding;
4. help preserve the subtier subcontractors and suppliers that are an essential element of the DTIB; and
5. keep costs under control.

Each of these issues is examined below.

[21] Michael I. Keller, consultant, personal communication, Mar. 16, 1992.

[22] Telephone interview with Gen. Philip L. Bolte (U.S. Army, ret.), former program manager, Bradley Fighting Vehicle System, Nov. 19, 1991.

[23] The requirement for this type of vehicle remained, however, and over a decade later the Army changed the name of the program and again requested bids for an Armored Gun System. The three companies invested additional funds and offered their prototypes in response to a new Request for Proposal (RFP). One company and its suppliers may be selected to produce this vehicle and recoup part of the prototype costs. If none of the three are selected, however, all will have lost not only their original prototype investment but the additional costs of upgrading and bidding again on the Army's RFP.

Preserving Design Teams

Design teams are important because the development of major weapon systems is as much an art as a science. Data alone cannot create a manufacturing capability; the other essential ingredients are people, infrastructure, knowledgeable management, and shop practices. As production budgets shrink, it will be essential to preserve the right design and manufacturing people to retain diverse approaches to defense systems work. Moreover, in order for design teams to be effective, they must work on real systems that may be actually built and tested.[24]

Preserving design teams means keeping them supplied with interesting and challenging work. **Given the new financial constraints, however, the number and size of design teams involved in prototyping will have to be reduced from the current level.** Over the past few decades, the increasing complexity of defense systems has led to the rapid expansion of design teams. At General Dynamics Convair Division, for example, the Tomahawk cruise-missile program started out with 8 to 10 people working on a small conceptual study and peaked at 300 to 400 engineers and other professionals at the start of EMD. At the Lockheed Skunk Works, the F-117 stealth fighter program involved a core team of about 300 throughout the development effort, but doubled in size to about 600 during the demonstration and validation phase. The EMD phase for a combat aircraft typically involves a staff of 3,000 to 7,000 people. On average, a fighter design team numbers about 1,000 people and costs about $100 million a year to maintain.

The current array of design teams will have to be consolidated into fewer, high-quality teams through streamlining, mergers, or strategic alliances. One approach is the "agile manufacturing" concept developed by the Iacocca Institute, which focuses on teaming arrangements. Companies form temporary consortia to bring together a critical mass of skills or resources for responding to a particular market opportunity, and then disband to restructure for the next demand.[25] There is no reason why Service laboratories could not participate in such teaming arrangements. National laboratories such as Lawrence Livermore and Los Alamos might also play an engineering support and training role.

In addition to cutting back the number of design teams, the size of the teams will need to be reduced. Two current trends should facilitate this process. First, **modern management systems, supported by advances in computer-aided design (CAD) and computer-aided manufacturing (CAM) technologies, can reduce the size of design teams by increasing the efficiency of the development process.** One example has been the development of new techniques for converting CAD models directly into three-dimensional hardware mockups (box 3-D). Another important advance has been the use of a single, integrated computer database to store all of the information needed to design, build, and support a weapon system. It might contain, for example, a geometric model of the more than 100,000 engineered parts that go into a combat aircraft, including cable runs, wiring harnesses, and hydraulic systems. Such models reduce the need to build expensive full-scale physical mockups to obtain insights into a system's appearance and internal layout.[26]

A centralized computer-integrated manufacturing (CIM) database can link together the functional departments of a company and its subcontractors and suppliers (figure 3-3). Since these different groups can work from the same information, it is possible to carry out a complex project with a smaller, more dispersed staff. Another advantage of an integrated database is that engineers can update the digital blueprints continually so that the latest version of the design is available to all users of the system. Moreover, design changes made at an engineer's desk can be communicated to a host of subcontractors in a matter of hours, rather than the days or weeks formerly required to print and mail them.

Although many integrated databases are still experimental, they have been used successfully for the development of the B-2 bomber, the YF-22 fighter, and the Boeing 777 commercial airliner.[27] It

---

[24] Paul H. Richanbach et al., *The Future of Military R&D: Towards a Flexible Acquisition Strategy*, IDA Paper P-2444 (Alexandria, VA: Institute for Defense Analyses, July 1990), p. 16.

[25] Roger Nagel and Rick Dove, *21st Century Manufacturing Enterprise Strategy* (Bethlehem, PA: Iacocca Institute, Lehigh University, 1991).

[26] In some cases, however, mockups still provide an economic way to determine hydraulic line runs, engine fit to fuselage, and maintenance requirements.

[27] "Computer System Design Reflects B-2's Complexity," *Aviation Week & Space Technology*, Nov. 28, 1988, pp. 26-27.

Box 3-D—Rapid Prototyping

Computer-driven tools are increasing the ability to move rapidly from designs to prototypes and thence to production. For example, a new technique known as *stereolithography* uses computer-aided design (CAD) data to produce three-dimensional solid models from a vat of photosensitive chemicals, which polymerize and solidify into plastic as they are irradiated with a laser beam. As a result, a design engineer can complete a design and produce an accurate physical model of a complex component in a single day, for technical and presentation mockups as well as prototypes. Quickly produced models of components can check fit against adjacent parts before expensive machining.

Stereolithography cuts the time needed to produce a mockup of a part by more than 90 percent.[1] For example, the Air Force Manufacturing Technology (MANTECH) Program used the technique to redesign the brake pedal on the B-52 bomber. A CAD representation of the redesigned pedal was converted by stereolithography into a plastic model, which was test-fitted into a B-52 cockpit. The pedal's dimensions were found to be incorrect, so the CAD design was modified and used to generate a second prototype, which fitted correctly. Turnaround time between discovery of the original design flaw to creation of the second prototype was about 7 days, a time savings of 6 to 8 weeks over conventional machining methods. According to the Air Force, the fact that the problem was identified and corrected early, before manufacturing began, yielded a substantial cost savings.[2]

The National Science Foundation and a group of private companies are currently supporting research to make rapid prototypes from CAD models using a full range of materials, from steel to ceramics. One approach involves using a printer nozzle to squirt a binder chemical onto a bed of powdered ceramic or metal, after which the part is solidified by firing in a furnace. This method can be used to produce solid parts, dies, or ceramic molds for metal casting. While technical obstacles remain to be overcome, this approach may eventually enable manufacturers to produce small lots of customized metal or ceramic parts directly from CAD models, without casting or machining.[3]

---

1 Alan S. Brown, "Rapid Prototyping: Parts Without Tools," *Aerospace America*, vol. 29, No. 8, August 1991, pp. 18-23.

2 "Rapid Prototyping Program Supports B-52 Brake Pedal Redesign," *USAF Manufacturing Technology Program Status Report* (Wright-Patterson Air Force Base, OH: Wright Laboratory MANTECH Program, December 1990), p. 5.

3 Gary Stix, "Desktop Artisans," *Scientific American*, vol. 266, No. 4, April 1992, pp. 141-2.

might be possible in the future to use computerized databases to develop a new design, upgrade it at regular intervals as new technologies become available, and build it when the need arises. Preliminary designs and manufacturing plans for such "mobilization prototypes" could be developed for rare contingencies such as Arctic warfare, special-operations needs, or future mobilization requirements.[28]

A second trend should also make it easier to rationalize the prototyping process. **Design teams are increasingly being restructured into multidisciplinary development teams that develop products and manufacturing processes simultaneously, an approach known as "concurrent engineering" or "integrated product development."**[29] Specialists are brought together at the beginning of the design process to exchange and define the information needed to manufacture and support the desired product. During development, this multidisciplinary team flows through multiple program assignments and is backed up with needed specialist support.[30] In the automobile industry, multidisciplinary development teams generally break up at the end of each development program. A prototyping-plus strategy, however, would seek to keep teams together between projects—an objective requiring a continuous flow of new prototyping projects. One approach would be to stagger prototyping efforts in time, so that some systems are in the conceptual design phase while others are in technology demonstration or limited production.

---

28 Leonard Sullivan, System Planning Corp., personal communication, Jan. 22, 1992.

29 Although concurrent engineering is the more common term, it is a misnomer because the process involves more than engineering.

30 Robert I. Winner et al., *The Role of Concurrent Engineering in Weapons System Acquisition*, IDA Report R-338 (Arlington, VA: Institute for Defense Analyses, December 1988), pp. 91-92.

Figure 3-3—Centralized Database Concept

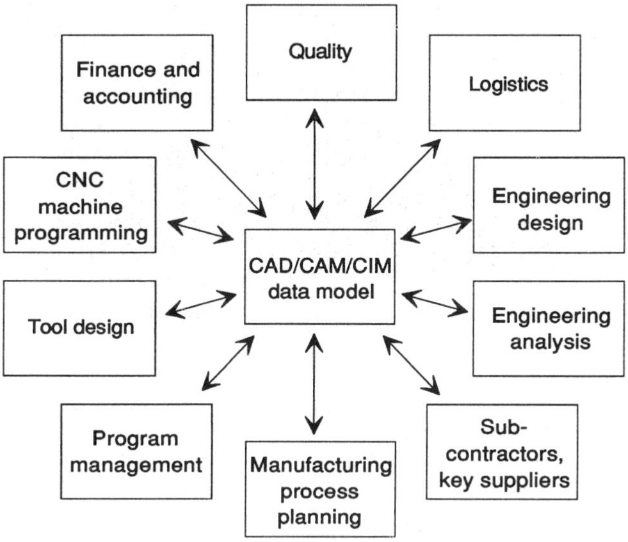

SOURCE: Office of Technology Assessment, 1992.

Maintaining Manufacturing Technology

Prototyping is a manufacturing activity—albeit one that differs from quantity production.[31] Technology demonstrators or advanced-development prototypes are usually built in special facilities, with little emphasis on durability, reproducibility, maintainability, or the suitability of the design for quantity production. Prototype construction is small-scale, flexible, and usually involves a small number of engineers or technicians working in stationary assembly booths or short, slow assembly lines. In contrast, an operational weapon system should be designed for efficient production on an assembly line and a long lifetime in the field. Quantity production is highly organized, requires a larger and more specialized workforce, and may entail the participation of several firms.

Given the different characteristics of prototype construction and quantity production, the transition from an advanced-development prototype to the final production item has traditionally been difficult and costly. In particular, it has been necessary to work out major bugs in the manufacturing process before production begins to run smoothly. For example, it took Martin Marietta 14 months to eliminate problems in the fabrication of its LANTIRN night-vision and targeting system. In the cases of the AMRAAM missile and the B-2 bomber, the transition from development to production has taken years. Industry officials argue that if they merely hand build a prototype or perform a limited production run, they will encounter serious problems in the transition to quantity production.

A possible solution to these problems lies with concurrent engineering, in which the design of a product and its manufacturing process are developed in parallel. By integrating manufacturing issues into the design process, concurrent engineering lowers the number of costly engineering changes needed after a system has entered production, significantly lowering total acquisition costs.[32] Boeing, for example, expects that concurrent engineering will reduce the development costs of its 777 passenger aircraft by as much as 20 percent.[33] For concurrent engineering to work, however, design and manufacturing engineers must share the same information. Organizational barriers must be broken down to permit the early release of preliminary design information to production staff and the feedback of manufacturing information to designers. Concurrent engineering is said to be "a people and communications issue, not an engineering technology one."[34]

The defense industry can learn from advanced civilian manufacturing in this area. Toyota and Honda, for example, make extensive use of prototypes to identify and solve design and manufacturing problems at an early stage of product development.[35] Some U.S. automobile companies have also implemented concurrent engineering on specific projects.

[31] The term "quantity production" is relative. Most defense products are built in small volumes compared with most mass-produced products.

[32] The traditional sequential approach to development results in the need to make many costly design changes before a system can be manufactured efficiently. During the full-scale development of the Bradley fighting vehicle, for example, FMC Corp. made a total of 60,000 engineering change orders costing an average of $2,000 each. See John A. Alic, "Computer-Assisted Everything? Techniques and Tools for Design and Production," manuscript, p. 20.

[33] Dori Jones Yang, "Boeing Knocks Down the Wall Between the Dreamers and the Doers," *Business Week*, Oct. 28, 1991, p. 120.

[34] Joseph T. Vesey, "Speed-to-Market Distinguishes the New Competitors," *Research-Technology Management*, vol. 34, No. 6, November-December 1991, p. 36.

[35] Kim B. Clark and Takahiro Fujimoto, *Product Development Performance: Strategy, Organization, and Management in the World Auto Industry* (Boston, MA: Harvard Business School Press, 1991), pp. 179-180.

Chrysler developed its new $55,000 Viper sportscar with an 85-person multidisciplinary development team, about a tenth the size of most U.S. automotive design teams. The team included 6 technicians who built all of the Viper prototypes. To transfer the manufacturing lessons learned from prototyping to production, the same 6 technicians were put in charge of assembly teams at the manufacturing plant, where 120 skilled production workers build the cars.[36] Since the Viper is a low-volume, high-value product that is largely hand assembled, it has much in common with defense systems like fighter aircraft.

Some defense contractors are beginning to address manufacturing and producibility issues during the demonstration and validation phase. In developing the X-31 demonstrator, for example, Rockwell International fielded a core multidisciplinary team of 50 to 60 design, manufacturing, and quality engineers who were retained throughout the various phases of the program. This approach resulted in better continuity of knowledge and institutional memory. Similarly, in developing the M1A2 tank prototype, the management of General Dynamics' Land Systems Division decided to have the prototype hardware built by workers in a production facility rather than by engineers in a specialized project shop. Although this approach initially sparked resistance, it promoted greater manufacturability by forcing designers and manufacturing engineers to work together.

The higher up-front costs associated with concurrent engineering are generally recouped during the production phase through a greatly reduced number of design changes and lower life-cycle costs. Nevertheless, the DoD has been reluctant to invest in manufacturing process development without a high probability of quantity production, even though some level of investment is warranted simply to maintain skills and improve manufacturing technologies. The dilemma is that whether a prototyping program will culminate in production is not usually known at the outset, because the decision depends on the outcome of the prototyping process itself. DoD and Service leaders must therefore weigh the early costs of concurrent engineering against its benefits in easing the potential transition to quantity production. Nevertheless, **even if only a small fraction of**

Photo credit: U.S. Army

Operational prototype of a M1A2 Abrams main battle tank undergoes field trials. An upgrade of the M1A1 tank, it has a better cannon, armor, electronics, and communications.

prototypes lead to a design that is produced in quantity, the savings achieved through concurrent engineering—and the concomitant benefits to the manufacturing technology base—may be great enough to warrant using this approach for most prototyping programs.

Alternatively, OSD and Service leaders could examine a prototype at multiple decision points during the development process and assess the probability that it will lead to a design that is produced in quantity. In this way, **the extent of investment in manufacturing process development and preproduction planning during prototype development could be calibrated to the probability that the system will enter quantity production.** Other factors that may influence the extent of investment in concurrent engineering during a prototyping effort include program goals, changes in the military threat, foreign technological advances, available funding, performance requirements, and acceptable levels of technological and financial uncertainty.

Some critics contend that a prototyping strategy would be incompatible with concurrent engineering. Ongoing advances in manufacturing, they argue, would render a finished but shelved design either obsolete or incompatible with new manufacturing processes by the time it entered production years later.[37] One way of addressing this problem would

[36] David Woodruff, "The Racy Viper is Already a Winner for Chrysler," *Business Week*, Nov. 4, 1991, p. 36.

[37] Donald Christiansen, "Design, Don't Build?" *IEEE Spectrum*, vol. 29, No. 3, March 1992, p. 23.

be for multidisciplinary design teams to update a prototype design periodically to keep up with significant improvements in product and process technologies. Limited production of selected prototypes would also make it possible to work out the major bugs in the manufacturing process.

Tooling is another important element of prototype construction. Fabrication and assembly tooling can be either "hard" or "soft," depending on its durability and the extent to which it is amenable to change. Hard tooling refers to metal dies and jigs that are sufficiently specialized, resistant, precise, and efficient to permit quantity production.[38] Soft tooling, in contrast, is designed for low-rate manufacturing and includes standard tools, improvised rigging and clamping, dies made of malleable materials such as zinc alloys, manual forming and welding processes, and the use of machined parts rather than precision forgings. In the automobile industry, for example, prototype body panels are formed slowly on soft dies, whereas production panels are stamped on high-speed, high-power press machines fitted with hard-metal dies.[39] Soft tooling is easier, faster, and cheaper to manufacture, but it is suitable only for short or low-rate production runs and results in greater variability in production.

With smaller U.S. forces and a reduced requirement for new weapons, it should be possible to rely more on soft tooling, which would be sufficient for low-volume production. For example, although Northrop and McDonnell Douglas built only two prototypes of the YF-23 fighter with soft tooling, they claim that they could have used the same tooling to manufacture 50 of these aircraft, or more than enough for field testing.

Soft tooling also provides the flexibility to modify a design from time to time. In future weapon systems, subsystems will be upgraded at regular intervals and structures may be modified; for example, the F-117 airframe was refined repeatedly to reduce its radar signature. As a result, more flexible tooling and frequent design changes may become the rule, not the exception. Given the expected declines in production over the next decade, industry could use prototype construction on soft tooling to solve manufacturing problems at an early stage, and to produce operational prototypes in limited quantities for field testing. **The challenge will be to build prototypes with soft tooling because of its flexibility and low cost, while simultaneously maintaining the capability to make a successful transition to hard tooling for quantity production.**

In the event of crisis or war, prototype production could continue on soft tooling while manufacturing engineers prepared the hard tooling required for quantity production. This approach is not new: between the World Wars, the United States developed 37 prototype tanks but produced none in quantity. After the outbreak of World War II, it took industry about 2 years to begin turning out large volumes of tanks. Although the more sophisticated weapon systems in today's arsenals would require a longer lead-time to reach high rates of production, **the length of time required for a major new threat to emerge would still provide enough warning to gear up production of major weapon platforms such as tanks and bombers.** Deputy Defense Secretary Donald Atwood has said, "We talk now of a warning time of a major land war in Europe of something like 1 to 3 years. That's plenty of time to reconstitute an entire new [industrial] plant, an entire new supplier base."[40] While short-warning regional conflicts would require a surge production capacity for munitions and other battlefield consumables, such wars would be fought mainly with forces-in-being. (See ch. 4.)

In the future, the definition of soft tooling may change as manufacturing systems become more versatile. Indeed, the long-term goal may be to increase the flexibility of manufacturing systems to the point where hard tooling becomes obsolete. It is already possible to download some types of CAD data to computer numerically controlled machine tools, so that a part can be designed and manufactured electronically without creating a paper drawing. Using this technique, it is possible to machine complex parts in 5 days, compared with the 40 days previously required. As computer-aided manufacturing technology matures, it should become possible to fabricate prototype components with the same machine tools as quantity production, to build prototypes on assembly lines designed for multiple products, and to achieve a rapid transition

[38] In the electronics industry, however, hard tooling refers primarily to specialized test equipment.

[39] Clark and Fujimoto, *Product Development Performance*, op. cit., footnote 35, p. 180.

[40] Atwood, quoted in "New DoD Weapon-Buying Approach Has Industry Crying 'Uncle'," *Armed Forces Journal International*, March 1992, p. 12.

from prototyping to quantity production. To date, however, neither military nor civil manufacturers have absorbed the most advanced production technology.

Limited production of prototypes would mean foregoing many of the cost-efficiencies that result from moving down a production learning curve. Nevertheless, some analysts argue that the largest gains in efficiency result from production of the early units, when the major bugs are worked out of the manufacturing process.[41] If this assumption is true, then **even the limited production of prototypes designed for manufacturability would significantly reduce the risks involved in the transition to quantity production.** In sum, greater use of concurrent engineering in prototyping and limited, intermittent production of prototypes for operational testing would help preserve key manufacturing skills while facilitating the transition to quantity production when necessary.

Preserving the Vendor Base

Even if prime contractors agree to prototype new systems, where will the necessary parts and components come from? Without full production lines, the number of subcontractors and component suppliers at the lower tiers of the DTIB may continue to erode, and skilled machinists and other manufacturing tradesmen may be lost. Thus, for a prototyping-plus strategy to work, the survival of the vendor base must be assured.

Given some ongoing production, there is no reason to expect that lower-tier suppliers will be reluctant to supply prime contractors developing system prototypes. Indeed, vendors often seek out such programs because it helps them to pursue their own advanced-technology development efforts. The United States will continue to have some production programs under way in most defense areas for the forseeable future, including low-rate production of current systems, overhauls, and retrofits. Throughout the 1990s, for example, production lines for three major combat aircraft may be active at any given time. While some industrial sectors such as tanks might be without production for a time, they are the exception rather than the rule.

To attract vendor participation, however, it will be necessary to reform the acquisition process. First, the government may have to provide substantial amounts of R&D funding and probably some guarantee of future military orders. Morever, vendors may refuse to accept R&D contracts because of the government's insistance on ownership of all technical data developed with public funding. (See ch. 2.) It will therefore be necessary to resolve the data-rights issue. Further, since many of the larger vendors sell primarily to commercial markets, convincing them to stay in the defense business may require modification of procurement regulations and military specifications.

In addition to these general approaches, there are some other options:

1. **The DoD could fund programs to retrofit and upgrade current platforms, ordering the improved components in sufficient quantities to make their development and manufacture profitable for subtier suppliers.** The government might also support, on a cost-plus basis, development of the tooling needed to manufacture essential components. Further, the DoD might pay prime contractors to integrate several new subsystems and components into technology demonstrators or advanced-development prototypes.
2. **The subtier base will need to be consolidated.** Prime contractors could protect their own workforce from layoffs by moving the production of key subsystems and components in-house. Alternatively, subtier firms might be consolidated into a smaller number of diversified companies, which would be linked to prime contractors through strategic alliances. Indeed, Total Quality Management (TQM) precepts call for the use of fewer, but high-quality and efficient, suppliers. Although market forces will result in consolidation, Federal Acquisition Regulations mandating "free and open competition" may need to be changed to permit long-term supply relationships between primes and subtiers.
3. **Subcontractors and suppliers might play a more active role in cooperating with prime contractors on prototype development and engineering.** For example, representatives of key suppliers and subcontractors might participate in concurrent engineering teams. This approach would broaden the training base and

[41] Linda Argote and Dennis Epple, "Learning Curves in Manufacturing," *Science*, vol. 247, No. 4945, Feb. 23, 1990, pp. 920-924.

improve timely response to emerging military requirements.

4. **The role of foreign suppliers should be considered.** Foreign sources often have a lock on subcomponent technologies (such as materials, semiconductors, and optics) that will be critical to any future systems. To ensure access to these technologies, the DoD may have to make defense contracts available to foreign vendors on more or less the same terms it offers domestic producers. Alternatively, the DoD could invest more money to develop or expand an onshore (North American) production capability, using Title III of the Defense Production Act.

Time and Cost of Prototyping

On average, an advanced-development aircraft prototype can be built for 25 to 30 percent of the total development cost of the system. But the actual cost of a prototyping-plus strategy would depend on several factors:

1. the type of system, desired military performance, and extent to which it is a radical departure from current systems;
2. thc number of contractors (and development teams) building prototypes;
3. the number and category of prototypes to be built (e.g., breadboard, brassboard, or fieldable operational prototype);
4. the amount of time a contractor is allowed for early development models;
5. the extent to which prototype design and manufacturing data must be documented for storage or recycling; and
6. the producibility of the design and its fidelity to the final production model, including the extent of systems integration.

Because of these numerous factors, **the cost and time involved in prototyping can vary enormously.** Whereas the 10 prototypes of the M1A2 tank (an upgrade of the current M1A1) are said to have cost about $15 million apiece, a radically new tank design (based on novel composite materials) could cost as much as $200 million. Similarly, while an austere technology demonstrator like the X-31 was developed under a cost-plus contract totaling about

Photo credit: DoD

X-31 technology demonstrator was unveiled in March 1990. Two of the aircraft were developed and built jointly by Rockwell and the German firm MBB.

$200 million (of which the U.S. share was $135 million),[42] four advanced-development prototypes of the Advanced Tactical Fighter (ATF) cost a total of about $5 billion to develop.

The cost of developing a prototype was not a major issue when it was just one step in a process culminating in quantity production. For tactical aircraft programs, for example, prototyping represented only a small percentage of total acquisition costs: the YF-16 prototype cost about $100 million out of a $30 billion program; the A-10 prototype cost about $100 million in a $5 billion program; and the AV-8B prototype cost $150 million out of a $10 billion program.[43] But the economics are very different when prototyping is no longer an integral step in a sequence leading to quantity production. **Without production to spread R&D and overhead costs over time, all equipment and associated costs must be borne during the development phase. The result will be an apparent rise in defense R&D costs.**

There are, however, some options for reducing prototyping costs. For technology demonstrators, one approach is to build unmanned, remotely operated systems that are easily reconfigurable. Whereas the safety requirements for human operators drive up costs, unmanned vehicles can provide

[42] Michael A. Dornheim, "X-31 Flight Tests to Explore Combat Agility to 70 Deg. AOA," *Aviation Week & Space Technology*, vol. 134, No. 10, Mar. 11, 1991, p. 38. The $200 million figure includes the design and construction of two prototypes and initial flight testing.

[43] Karen W. Tyson et al., *Acquiring Major Systems*, op. cit., footnote 15, p. VIII-2.

useful information at no risk to human life. (See box 3-E.) In the case of advanced-development prototypes, costs can be reduced by building subscale models when the effects of scale are understood. During the development of the Avro Vulcan strategic bomber, for example, the British saved money by building two full-scale prototypes to evaluate flight characteristics, and four subscale prototypes to test other aspects of the aircraft such as power-control systems and electronics.[44] Another approach is to prototype only the critical components of a weapon system. In prototyping an aircraft carrier, it might be sufficient to build a control tower on a barge to test the command-and-control, threat-assessment, and other systems.

Finally, the United States might consider engaging in more collaborative prototyping programs with the NATO allies, the industrialized countries of the Pacific Rim, and possibly Russia. The advantage of international collaboration is that it permits sharing of development costs and enables U.S. firms to gain access to foreign technologies. **Collaboration is likely to become a more attractive option as defense budgets are reduced and U.S. forces engage in multinational military operations, such as the Gulf War, reinforcing the need for interoperability.** A drawback is that collaboration can increase U.S. dependence on offshore sources; it also inevitably entails compromises on program objectives, specifications, schedules, and worksharing. Further, transfers of U.S. technology might enable some foreign firms to become more formidable competitors in the future. **Collaborative programs must therefore provide for a two-way flow of important technologies, so that the U.S. industry gains at least as much as it gives.**

Although West European firms have engaged in joint development programs since the mid-1950s, this approach is relatively new for the United States. A recent example is the X-31 technology demonstrator, jointly developed by Rockwell International and the German firm Messerschmitt-Bölkow-Blohm (MBB). Launched in 1986, this program was funded under the 1985 Nunn-Quayle Amendment. According to a Memorandum of Understanding between the U.S. and German governments, the X-31 program is managed jointly by DARPA as overall program manager and the German Ministry of Defense as deputy program manager. The development and production work has been divided between the two firms in proportion to each country's financial contribution to the program (about 72 percent American and 28 percent German), and a joint working group resolves all interface problems.

The collaboration has worked well because Rockwell and MBB have complementary technological strengths. Whereas MBB developed the basic enhanced-maneuverability concept, Rockwell offered its system-integration skills. Both firms benefitted from sharing resources, people, and ideas. Based on their positive experience with the X-31, MBB and Rockwell plan to collaborate on other projects. In addition, the U.S. and German goverments are considering a 5-year joint research program aimed at making fighter

Box 3-E—Remotely Piloted Research Vehicles

To explore the limits of fighter maneuverability achievable with current structural and propulsion technologies, NASA and the Air Force contracted Rockwell International to develop a remotely piloted research vehicle called Highly Maneuverable Aircraft Technology (HiMAT). The HiMAT contained a TV control system, telemetry, and a suite of research instruments. Because of its modular design and construction, the basic components of the aircraft could be altered to evaluate design changes, such as new relationships among control surfaces, modified airfoils, and various types of thrust-vectoring engine nozzles.[1] Other advantages of the HiMAT vehicle were its reduced size, which made it inexpensive to build and operate, and the fact that it could withstand accelerations that would kill human pilots. The chief drawback of the system was the need to develop a parallel command-and-control structure on the ground to operate it.

DARPA has developed a related concept known as Advanced Configuration Remotely Operated Basic Agility Technologies (ACROBAT), a family of subscale demonstrator aircraft that would be flown remotely from a computer terminal on the ground and whose configuration could be easily modified.[2]

[1] Christopher Chant, *Aircraft Prototypes* (Seacaucus, NJ: Chartwell Books, 1990), p. 118.

[2] Interview with Lt. Col. Michael S. Francis, Advanced Systems Technology Office, DARPA, Sept. 1991.

[44] Chant, *Aircraft Prototypes*, op. cit., footnote 9, p. 32.

Photo credit: DoD

Remotely operated experimental aircraft called HiMAT was designed to test new technologies for future fighters. Less expensive than a manned aircraft, it can do high-G maneuvers without risking pilots' lives.

aircraft more maneuverable, building on the results of the X-31 demonstrator.[45]

Rethinking the Acquisition Process

Throughout the cold war, the defense industry was oriented toward the need to counter a large and immediate Soviet threat. But the waning of that threat has given the United States the opportunity to shift its emphasis from short-term military capabilities to long-term military potential. In the new security environment, developing multiple prototypes makes more sense than committing scarce resources to the production of current-generation weapons, of which there is already an abundance. A prototyping-plus strategy would provide an opportunity to continue technological innovation, maintain the defense technology base, and prepare for the future. It would also keep design teams together and, through the judicious use of concurrent engineering and limited production, help to maintain manufacturing skills.

To hedge against uncertainties in both technology and the security environment, the number of prototyping programs should be large relative to the number of systems that enter quantity production. Even though most prototyping programs would not lead to a design that is produced in quantity, they would still yield useful information and technologies that could be recycled into the next generation of systems or transferred to other programs. Since prototyping is a form of experimentation, it would not be redundant to build multiple prototypes with dissimilar designs in response to a given military requirement.

Shifting to a prototyping-plus strategy would entail a fundamental "cultural" change in both the defense industry and the government weapons-acquisition community. First, it would require a restructuring of the weapons-acquisition process away from the linear pipeline process culminating in production. The current model would be replaced with a new paradigm in which prototypes are developed to acquire new technical knowledge and to enhance the Nation's long-term readiness against a spectrum of possible threats. Selected prototypes would be manufactured in limited numbers on soft tooling for operational testing; when a military requirement arose, prototypes could be moved into quantity production.

Although greater use of concurrent engineering would reduce development time and total procurement costs, the DoD must give defense contractors incentives to develop more manufacturable systems. One approach would be for the DoD to award prototyping contracts based on the performance, manufacturability, and maintainability of proposed designs. The winning firm might also receive the added bonus of a contract for limited production of the prototype, without second-source competition. At the same time, it will be necessary to discipline the development process with cost and schedule targets; otherwise, designers will never stop tinkering, and no one at the user or procurement level will abandon the quest for the ideal solution. A streamlined approach to development, known as quick-reaction prototyping, was used successfully during the Gulf War. (See box 3-F.)

A prototyping-plus strategy would also require restructuring the defense industry to reduce capacity and create more flexible manufacturing practices, such as multiproduct assembly lines. To this end, the DoD would need to support the development of innovative manufacturing processes and novel materials, such as the radar-absorbing composites used in stealth aircraft. This investment would be critical because the very nature of most defense production—uncertainty over orders, the

[45] Barbara Opall, "U.S., Germans Plan Research on Fighter Jets," *Defense News*, vol. 7, No. 4, Jan. 27, 1992, p. 4.

Box 3-F—Quick-Reaction Prototyping

During the Persian Gulf War, personnel from Texas Instruments, Lockheed Missiles and Space, and Eglin Air Force Base took only 37 days to develop the GBU-28 penetrator bomb, which was then used to destroy an Iraqi command bunker that had survived direct hits from 2,000-pound bombs. Development of the new weapon required great speed and secrecy, use of existing industrial capacity and parts, and cooperation among private firms, Army arsenals, and an Air Force base.

Development of the GBU-28 began on January 21, 1991, in the midst of the air campaign against Iraq. The Air Force gave industry and its own designated project staff a free hand to get the job done as quickly as possible, with a minimum of red tape. As a first step, Eglin personnel requested the use of old 8-inch howitzer barrels stored at Letterkenny Arsenal in Pennsylvania. The gun barrels were shipped to Watervliet Arsenal in New York, where they were machined into the bodies of the new bombs. Lockheed then developed the warhead, while Texas Instruments developed the guidance units. Designers at Texas Instruments took only 4 days to craft a quarter-scale aluminum model of the bomb for wind-tunnel testing of the body and tail-fin configuration.

Meanwhile, other TI engineers used computer simulation to develop guidance software for delivering the bomb with pinpoint accuracy. The TI team compressed the software development and testing—normally an 18-month to 2-year process—into less than 2 weeks. After field-testing at ranges in Nevada and New Mexico, two GBU-28s—each more than 18 feet long and weighing 4,700 pounds—were flown to Saudi Arabia. They were then fitted to the undercarriages of a pair of F-111s and used successfully on February 27, 1991, to destroy the Iraqi command bunker at Al Taji Air Base north of Baghdad.[1]

[1] Gregg Jones, "Genesis of a Bomb: TI's Role Critical in Quick Development of Weapon," *Dallas Morning News*, June 30, 1991.

small size of production runs, excess capacity and the consequent difficulty in recovering the investment in tooling—makes defense firms reluctant to invest their own money to develop new manufacturing technologies.

Finally, a prototyping-plus strategy would require new approaches to program management. Specific changes might include new systems for monitoring costs, schedule, and performance; improved liaison with system users; new arrangements for managing subcontracts; and enhanced logistics planning to maintain the currency of prototypes. Other options for managing prototyping programs follow:

1. *Use performance criteria rather than specifications.* Giving prototype designers greater flexibility would enable them to trade off performance against cost. Cost discipline could be maintained through competition between prototype designs, government auditing, and positive fee or profit incentives for completing prototype development on time and under budget. In this way, contractors would have the freedom to be creative without having to give up proprietary information to "level the playing field."
2. *Reconsider the role of competition.* It would not make sense for every new prototype to undergo the 2-4 year source-selection process now used for most full-cycle procurement programs. Thus, competition may have to be achieved in more flexible ways.

POLICY IMPLICATIONS

Congress must decide whether it wishes to invest in maintaining innovation, preserving the defense technology base, and hedging against future technological breakthroughs by potential adversaries. If so, then a prototyping-plus strategy should be part of the answer. Since private industry will be unable and unwilling to invest its own money in prototype development without the immediate prospect of a lucrative production contract, the DoD will have to bear the full cost of prototyping. **Thus, for a prototyping-plus strategy to be viable, it would require a long-term funding commitment from Congress. Even so, the total cost would be considerably less than the alternative of maintaining a warm production base for most military items, which is simply not feasible in the current budgetary or strategic environment.**

There are several options for carrying out prototyping programs, including competition among private firms; sole-source development in public or private "arsenals;" and the use of specialized engineering firms ("design houses"). Industry officials contend that prototyping in public arsenals would not be effective because the government does not have a good track record as a systems integrator and would not face the same cost discipline as firms competing in the marketplace. The aerospace and armored vehicle industries also oppose the use of specialized design houses, although the Navy makes extensive use of them. Since design houses are less capable of concurrent engineering, they would result in higher downstream production and life-cycle costs. Moreover, without manufacturing experience, the transition from prototyping to production would be very difficult.

Alternatively, the DoD could award prototying contracts to full-service firms that do both R&D and manufacturing. Such firms might build prototypes on flexible production lines. (See ch. 4.) Another option would be to consolidate development and manufacturing in several Skunk Works-like organizations, which would build competing prototypes during the concept-definition phase. Advocates of this approach argue that it would promote fresh technological approaches and force efficiencies through competition.

Other questions about a prototyping-plus strategy remain to be answered. How can prototyping contracts be made sufficiently interesting and profitable to motivate companies, scientists, and engineers to focus on state-of-the-art developments unique to military systems? With reduced defense budgets, how many prototyping-plus programs could be financed at any one time? How much would companies learn about manufacturing by using soft tooling? And should government laboratories assume the role of developing enabling technologies in those areas where the specialized nature of the application limits private-sector incentives? These unanswered questions suggest that while prototyping-plus is a promising approach, it will need more refinement before it is ready for implementation.

Finally, while a prototyping-plus strategy would preserve essential design and manufacturing capabilities and foster technological innovation, it could not by itself maintain the defense manufacturing base over time. Firms might rely on prototyping to preserve their core competencies, but they could only survive financially by: eliminating excess capacity; drawing on other businesses, such as supporting and upgrading fielded weapon systems; and diversifying into civilian markets. Prototyping-plus must therefore be seen in the context of a broad restructuring of the DTIB.

Chapter 4

Efficient, Responsive, Mobilizable Production

Contents

Boxes

Figures

Tables

Chapter 4
Efficient, Responsive, Mobilizable Production

INTRODUCTION

By dollar expenditure, production is the single largest component of the defense technology and industrial base (DTIB). Production will probably suffer the largest defense budget cut in absolute and relative terms. Historically, this component has had three principal functions:

1. manufacturing high-quality military equipment in peacetime,
2. responding quickly but selectively to increased military requirements in crisis or war, and
3. mobilizing the national economy for large-scale hostilities.[1]

Redesigning Defense suggested that these functions remain desirable characteristics for the future smaller production base. This chapter discusses options the Administration and Congress might employ to arrive at a future production base that is efficient, responsive, and mobilizable under the conditions of significantly reduced procurement.

BACKGROUND

Defense procurement is projected to fall over 50 percent in real terms between fiscal years 1985 and 1997. Between 1990 and 1993, budget authority for aviation is projected to decline by 40 percent, shipbuilding by 59 percent, and Army tracked vehicles and weapons (excluding missiles) by 77 percent.[2] While production of some munitions and other consumables may increase temporarily to replenish stocks consumed during the Persian Gulf War, procurement of major weapon platforms will decline sharply over the next decade.

Procurement reductions of this magnitude will radically change the way defense manufacturing is conducted. These reductions might severely weaken the defense production base if they are handled without sufficient foresight. The Nation may be hard pressed to maintain future shipbuilding, aircraft manufacturing, and armored vehicle production capabilities, for example. Small companies that produce critical components for major defense systems may become economically unviable and leave the defense business or cease operations entirely. And basic material and subcomponent suppliers may decide that the defense market has grown too small and unpredictable to be worth the trouble of dealing with procurement laws and regulations. (See ch. 6.) In order to survive the cutbacks and remain competitive, businesses may jettison important capabilities (e.g., R&D facilities and staffs) and put off new productivity investments. The end result might be the unnecessary loss of skilled workers and an inadequate DTIB.

As procurement authorizations declined in the wake of the Carter-Reagan military build-up, the production base was left with significant overcapacity in most industrial sectors. Reduced production, large overhead, and sunk costs caused weapon systems to grow more expensive even as the contractor and supplier base shrank. The decreasing global competitiveness of the U.S. economy made the military more dependent on foreign suppliers in such market segments as advanced materials, electronics, and display technologies. The projected future decline in defense procurement is expected to aggravate all of these trends.

In *Redesigning Defense*, OTA outlined three desirable characteristics for the future defense production base:

1. limited, efficient peacetime production capabilities for high-quality materiel;
2. responsive production of ammunition, spares, and consumables for theater conflict; and
3. healthy, mobilizable civilian production capacity.

Managing the transition to such a production base while avoiding the pitfalls of recent trends will require leadership from both the Administration and Congress. If any meaningful resolution of the dilemma in defense production is to be found, it will

[1] U.S. Congress, Office of Technology Assessment, *Redesigning Defense: Planning the Transition to the Future U.S. Defense Industrial Base*, OTA-ISC-500 (Washington, DC: U.S. Government Printing Office, July 1991), p. 3. Much of the introductory comments for this chapter are taken from this report.

[2] Steven Kosiak and Paul Taibl, *Analysis of the Fiscal Year 1993 Defense Budget Request* (Washington, DC: Defense Budget Project, Mar. 11, 1992), tables 8 and 9.

be necessary to focus on the end goal—a restructured defense industry.

THE CURRENT PRODUCTION BASE

The production base is not a monolithic structure amenable to generic remedies. It is a complex conglomeration of separate ventures on multiple tiers in many industrial sectors, with varying degrees of private and public ownership, operating in an environment of increasing global economic interdependence. The defense downturn will affect individual businesses differently, and effective solutions to the problems of the future production base will depend on understanding these differences. The current production base was described in some detail in *Redesigning Defense* and is only summarized here.

Tiers of the Base

The DTIB can be divided into a series of levels or tiers. Occupying the top tier of the defense industrial base are the prime contractors,[3] often large corporations (e.g., General Dynamics) whose main task is to bring together all the necessary components for a system and integrate them into a whole (e.g., an aircraft).

The vast majority of production base companies, however, are in the subtiers.[4] The subcontractor tier of the defense production base is the most diverse in terms of size and product, and includes both industrial giants and small machine shops. A subcontractor manufactures specialized parts, components, or subsystems that are integrated into a larger subsystem or final system. In a major weapon system, several layers of subcontractors might produce hundreds or thousands of individual items. The supplier tier provides the prime contractor and subcontractors with basic parts, hardware, subcomponents, capital equipment, and materials. This tier is generally more integrated into the civilian market than the prime or subcontractor tiers, although cases of suppliers totally dedicated to defense work are not uncommon. Figure 4-1 illustrates this multilayered arrangement for the DDG-51 *Arleigh Burke* destroyer.

Each tier of the base is already being adversely affected by the downsizing of the defense production base. **Prime contractors** are heavily dependent on large weapon system contracts, which are increasingly scarce. Many still have sufficient working capital from production contracts that began in the 1980s, since money appropriated is only now being spent.[5] This capital will allow some of them to reorient their business horizontally, to other markets (e.g., through acquisitions of defense and nondefense firms),[6] or vertically, by taking over the business of their subcontractors and suppliers. As current production contracts are completed, however, money will become increasingly scarce. Many prime contractors hope to expand sales of systems, repairs, spare parts, or upgrades abroad.

Larger, more diversified **subcontractors** should not be devastated by the termination of any single program. Most have substantial commercial dealings to help them weather defense cutbacks or allow them to leave defense work for the civil sector while their less diversified defense competition fails. For example, Allied Signal manufactures a wide variety of aerospace power systems, guidance systems, torpedo propulsion systems, sonars, and other electronics for the Department of Defense (DoD). It also does extensive work in commercial aerospace, as well as in the automotive and material sectors.[7] Like some of the primes, larger subcontractors are often

[3] The breakdown of the base into tiers (primes, subcontractors, and suppliers) is an artificial construct used widely to simplify discussion of the base. The actual base is more complex. For example, a major corporation may serve as prime contractor on one contract while acting as subcontractor on another, or a small company that functions as a prime contractor on a small item (e.g., shoes) may have characteristics more in common with subcontractors than a major prime contractor. These distinctions will be addressed in the text where important. For a further discussion on the tier structure see *Redesigning Defense*, op. cit., footnote 1, pp. 40-44.

[4] Over 70 percent according to the DoD, Defense Systems Management College, *Defense Manufacturing Management: Guide for Program Managers*, 3d ed. (Washington, DC: U.S. Government Printing Office, April 1989), ch. 2, p. 5.

[5] Major systems take years to build. In some cases, multiple buys, authorized and contracted for in one year, will be started over a period of several years.

[6] These companies prefer to acquire businesses that have large back orders or good commercial prospects. ("Casualties of Peace," *Business Week*, Jan. 13, 1992, p. 64.) For example, Hughes Aircraft plans to increase its proportion of commercial sales from 25 percent in 1988 to 50 percent by the late 1990s through investments in areas such as satellites, head-up displays, and electric drives for cars. (Caleb Baker, "Hughes Braves Skeptics With Commercial Market Drive," *Defense News*, vol. 6, No. 46, Nov. 25, 1991, p. 24.)

[7] "Top 20 Government Contractors," *Government Executive*, vol. 23, No. 8, August 1991, p. 119; and Dialog Information Services, Inc.

Figure 4-1—Production Tiers for the Navy's *Arleigh Burke* Guided Missile Destroyer

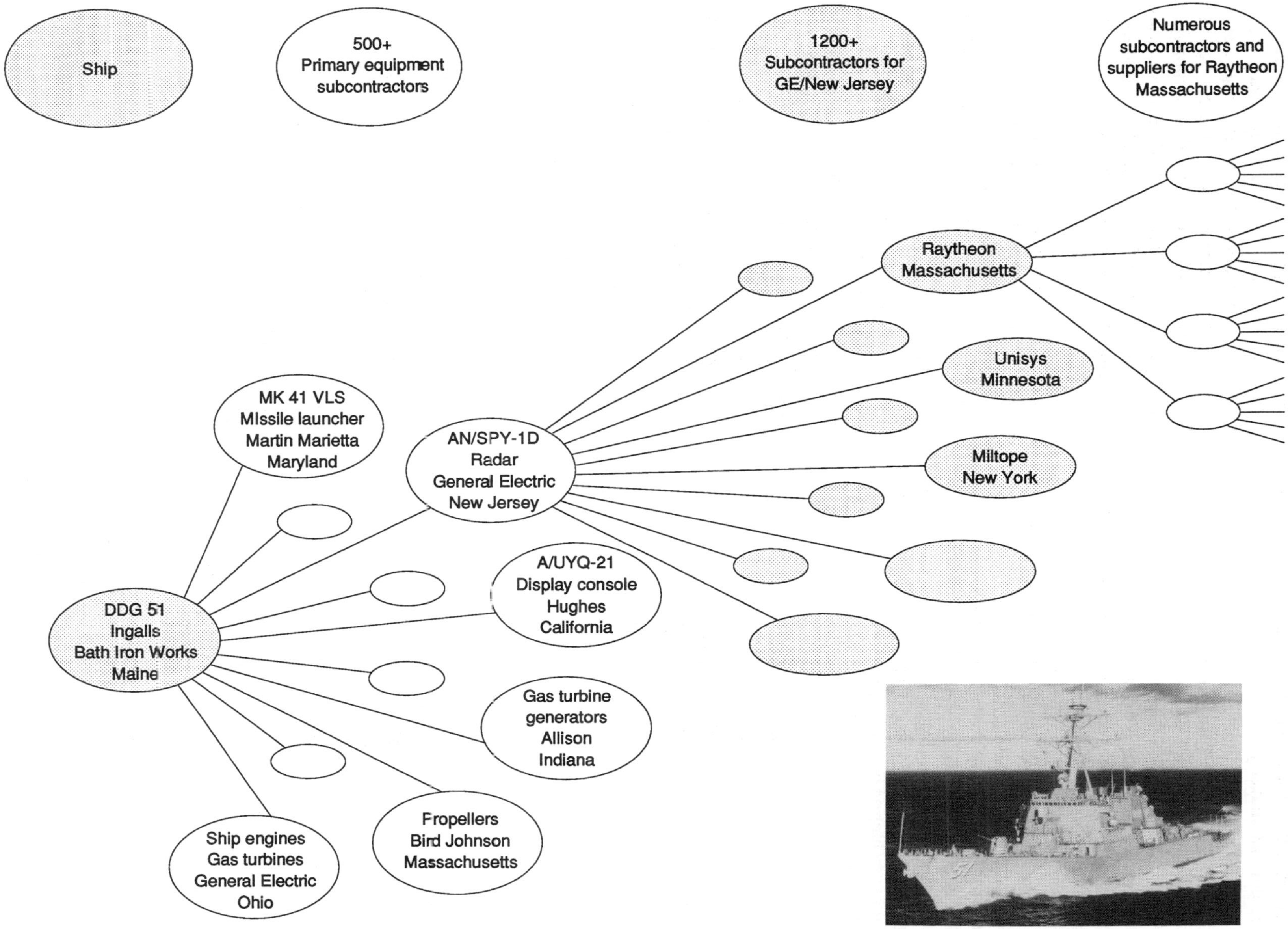

SOURCE: Naval Sea Systems Command, July 1990.

Photo credit: Department of Defense

able to acquire the necessary resources to expand vertically or horizontally into other markets.

Smaller subcontractors involved in only a few programs are more immediately at risk from reduced defense procurement. The elimination, delay, or stretch-out of programs could force them out of the defense business and into either the commercial world or bankruptcy. Many of these companies have made their living through the ability to meet unique military specifications and operate according to military auditing practices. The transition to commercial markets, standards, and practices will be difficult. Some subcontractors (e.g., electronic equipment manufacturers), both small and large, see a continuing need for their products, whether they are used as components in new systems or as upgrades in older military equipment, and thus are somewhat optimistic about their future.

The future health of many defense **suppliers** depends on their strength in the civil sector rather than on the future course of defense procurement, because their defense market is relatively small compared to their civil market. For example, the military's share of the domestic market for DRAM (dynamic random access memory) chips, which are used in a wide array of electronic devices, is only a few percent.[8] As defense cutbacks make this portion of the market even smaller, suppliers may find the stringent specification, handling, and accounting rules of government procurement increasingly burdensome. The result may be to force the DoD toward higher unit costs, commercial standards, or the creation of dedicated government suppliers. Suppliers that are more dependent on defense spending and regulations will face a fate similar to that of the less diversified subcontractors.

Outside the domestic defense production base, but intertwined with it, is **the global DTIB.** The DoD and its contractors routinely purchase materials, parts, components, and finished goods from foreign manufacturers, just as other nations do from the United States. Foreign militaries are a significant market for U.S. defense products. Through foreign sales, the United States is able to reduce unit costs on weapons and equipment and keep production lines warm when domestic requirements wane. There is, however, public concern over such sales.

Foreign defense production also supplements the U.S. defense base by sharing technology and processes through cooperative ventures, thereby reducing duplication of R&D, production, and maintenance. Foreign firms also sell components and materials that are either not available on the U.S. market or are less expensive. Such trade carries risks: shared technology could undermine domestic industry and foreign supplies could be cut off. But without this cooperation, the United States might not have access to some state-of-the-art militarily unique and dual-use technologies and would have to pay the cost of pursuing them independently or not having access to them at all.

Public and Private Sectors

The current production base is divided between the private and public sectors. The United States relies primarily on private industry to provide defense materiel. Most defense work is done at privately owned facilities. However, when the initial capital investment costs of a defense program are prohibitively high or when the government wants the option of shifting contracts among firms without having to reinvest in new infrastructure, the government may establish a government-owned, contractor-operated (GOCO) facility. The DoD owns a number of GOCOs, including aircraft assembly facilities, propellant and explosive plants, and tank production lines, which are run by private firms. The government also retains a few government-owned, government-operated (GOGO) facilities for assured access or to meet a requirement that the private sector is not fulfilling at a reasonable cost (e.g., large-bore gun tube production at the Watervliet U.S. Army Arsenal). Government ownership and operation provides the most direct government control over facilities and resources. Critics of GOGOs argue that private management is more efficient and innovative. Recent government policy has been to divest government holdings.

As defense procurement shrinks, it is likely that some unique subcontractors or suppliers of items critical to a weapon system will face business failure, threatening a shutdown in system production. The DoD will then have the choice of assisting the failing firm through higher prices, subsidies, or the purchase of facilities (making them GOCOs);

[8] Interview with Martin Libicki, National Defense University; and Benjamin Zycher, Kenneth A. Solomon, and Loren Yager, "An 'Adequate Insurance' Approach to Critical Dependencies of the Department of Defense," R-3880-DARPA, The Rand Corp., 1991, p. 23.

stimulating other sources (foreign or domestic) of production; redesigning the relevant weapon to bypass the missing component; or establishing a public production capability (i.e., a GOGO).

Representative Industries

The current defense production base is a heterogeneous collection of industrial sectors, which will be affected by procurement reductions in different ways. The following are examples of important industrial sectors.

Defense Electronics

Defense electronics appears to be the industry segment best positioned for the restructuring of the defense industrial base that lies ahead. Defense electronics firms are generally subcontractors on major system projects, although in some areas, such as command, control, and communications, the electronics firms assume the role of prime contractor. The larger firms tend to have several defense contracts under way at any one time. Although many electronics suppliers participate in the larger commercial electronics sector, strict military specifications and accounting procedures compel most firms to segregate civil from military production. The rapidly growing commercial electronics industry may provide companies fertile ground for horizontal expansion.[9] However, prospective commercial partners might shy away from long-term relationships with defense electronics firms for fear of being abandoned at the first upturn in defense procurement.

Defense electronic firms, while bracing themselves for cutbacks, see future opportunities as well. Even without the acquisition of major weapon systems—the bread and butter for the large prime contractors—electronic firms see upgrades of their products as inevitable because of the fast-paced development cycles in the world electronics market. Moreover, they foresee a continuing opportunity to supply electronic upgrades to foreign countries that have purchased American weapon systems in the past. In fact, new weapon system production will continue, albeit at a greatly reduced rate. When combined with upgrade and other programs, this production will eventually halt the downward trend and may even provide for moderate growth of defense spending in this sector. (See figure 4-2.) Spares and repairs are seen as less viable options for future business because of increased product reliability.

Photo credit: Lockheed Electronics Co.

Lockheed technician tests microelectronic components to ensure they meet military quality standards.

Satellites

The satellite industry is closely related to, and often intertwined with, defense electronics, especially at subtier levels. There are only a few major prime contractors. Like defense electronics firms, these firms tend to work on several projects at once, making them less dependent on a particular project. The main difference between the sectors is that satellites are generally built in small, high-value lots of one or a few at a time. This might make the prime contractors vulnerable should funding for satellites diminish. The satellite sector is hoping for increased commercial business and NASA construction as well as continued work on Strategic Defense Initiative projects such as the "Brilliant Pebbles" anti-

[9] For example, while the DoD demand for semiconductors is likely to grow at 2 to 3 percent annually, the commercial market is expected to expand at a rate of 13 to 15 percent. (Debra Polsky, "Chip Producers Turn Attention From Military to Boost Revenues," *Defense News*, June 10, 1991, p. 55.)

Figure 4-2—Projected Defense Electronics Procurement Budget Through 2001

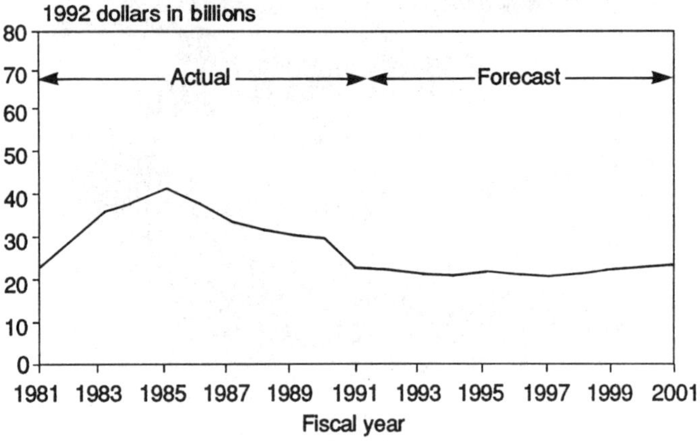

SOURCE: Electronic Industries Association, 1991.

Photo credit: The DoD

Troops prepare to board a UH-60 Blackhawk during Operation Desert Shield.

missile system or on the development of less capable but more numerous military satellites dubbed "light sats" or "cheap sats." But there is growing concern over foreign competition.

Fixed-Wing Aircraft

The military aircraft industry anticipates program cancellations, delays, and stretch-outs. Too many companies are chasing too few contracts. Industry analysts believe that the military cannot support the current number of aircraft prime contractors and that consolidation will be unavoidable. A Rand Corp. official, for example, predicted that the number of military aircraft divisions of major U.S. airframe manufacturers will shrink from 10 to 5 or fewer in the next 3 to 5 years through mergers, changes in organizational status, or leaving the business.[10] Companies are laying off or not replacing employees, closing or selling off excess facilities, and entering into teaming arrangements with their competitors to share both the risks and rewards of new contracts. Global competition in the military and commercial aviation business is intensifying with many foreign competitors buoyed by government subsidies, and foreign sales are increasingly subject to offset agreements that transfer technology to future competitors. The Air Force's F-22 Superstar interceptor and the Navy's AX attack plane appear on track for development and production. Continued production of some current models is also scheduled.

Helicopters

The U.S. military helicopter industry includes four major prime contractors, all of them divisions of major corporations. Military sales dominate U.S. production (more than 85 percent between fiscal years 1987 and 1990), but sales in the commercial sector are significant. In addition to extensive defense procurement cutbacks, the U.S. helicopter industry faces the possibility that the Army will transfer 3,000 aging helicopters into the commercial sector during the next decade.[11] Such surplus helicopters may further depress the demand for new commercial helicopters.[12] On the other hand, they may increase the demand for spare parts, upgrades, and overhauls.

In the 1960s, U.S. firms dominated world helicopter sales, only to be challenged in the 1980s by the emergence of aggressive foreign competitors, most of which are partially government owned or subsidized. Government support may give foreign companies an advantage over U.S. firms weakened by the reduction of military contracts, which have traditionally driven U.S. helicopter innovation. The

[10] Bruce D. Smith, "Airframe Building Capability Loss Looms for Full-Service Defense Contractors," *Aviation Week and Space Technology*, vol. 136, No. 11, Mar. 16, 1992, p. 41.

[11] Leonard M. Horner, testimony before the House Armed Services Committee Panel on the Structure of the U.S. Defense Industrial Base, Nov. 1, 1991.

[12] They may also undermine foreign military sales if they are passed on to allies.

Figure 4-3—Impact of Comanche Procurement on Military Helicopter Funding and Production

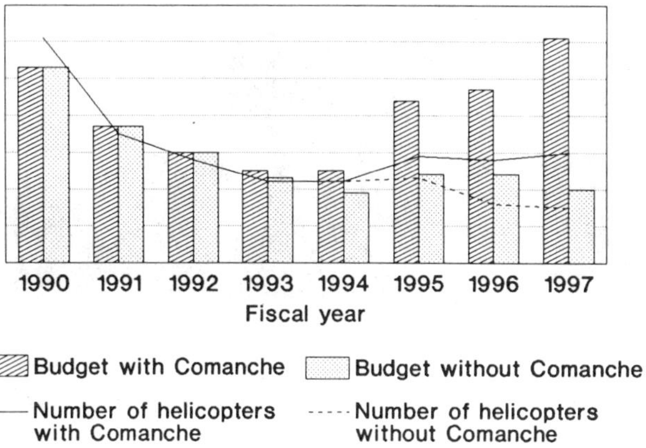

NOTE: Figure does not include V-22 *Osprey*.

SOURCE: DoD FY 1992-97 Program Objective Memorandum.

Photo credit: U.S. Army

The FMC Bradley armored fighting vehicle provides transportation, protection, and firepower for mechanized infantry, armored cavalry, and scout units.

future commercial helicopter market will likely be dominated by a competition to capture market share in other countries.

Projections of military helicopter production vary substantially depending on the systems built. For example, figure 4-3 illustrates the effect a decision to purchase the RAH-66 Comanche helicopter would have on procurement levels.[13] Because of the general wear and tear on helicopters, the need for repairs, upgrades, and spare parts should keep a core of subtier firms in business.

Armored Combat Vehicles

The Army is currently reviewing its plans for manufacturing armored combat vehicles. The diminished threat of large-scale conventional hostilities in Europe, the signing of the Conventional Armed Forces in Europe Treaty, the impressive performance of current armored vehicles in the Persian Gulf War, and projected budget reductions have left the Army with a large supply of advanced armored vehicles and an overcapacity for production. The Army had planned to phase out production of current combat vehicles and begin the development of a new family of six armored vehicles under the Armored Systems Modernization program. It now appears, however, that this family will be restructured around three vehicles, with the other three deferred indefinitely.

Reductions in Army procurement will have a substantial but varying impact on companies involved in producing armored vehicles. The two main armored vehicle systems, the Abrams tank and the Bradley fighting vehicle, have respectively over 1,000 and 200 subcontractors and suppliers, with relatively little overlap between the programs at the higher tiers. (See table 4-1.) The prime contractors for these systems—General Dynamics Land Systems for the Abrams and FMC for the Bradley—argue that unit costs may become unaffordable unless specific levels of production are maintained.[14] The DoD has stated that procurement of these systems will cease, leaving export sales, spare parts, and R&D on follow-on systems as the main tasks for the armored vehicle sector in the 1990s. Mothballing some facilities is seen as more cost effective than continued production.[15] While the current primes have considerable expertise in devel-

[13] The President's fiscal year 1993 budget request for the DoD emphasizes continued Comanche development and prototyping over production. Upgraded Apache and other helicopters and unmanned aerial vehicles are intended to fulfill the Comanche's role in the near term. See U.S. Congress, House Armed Services Committee, "Statement of the Secretary of Defense Dick Cheney in Connection with the FY 1993 Budget for the Department of Defense," Feb. 6, 1992.

[14] The minimum economic production rate for a particular plant is determined by a number of physical and organizational factors, as well by the measures taken at the plant to reduce overcapacity. Both General Dynamics Land Systems Division and FMC have taken significant steps in recent years to reduce their overcapacity and establish lower economical production rates. Government actions, discussed later, can further lower these rates.

[15] Department of Defense, *Report to Congress on the Defense Industrial Base*, November 1991, p. ES-5.

Table 4-1—Sample Products of the Armored Vehicle Production Tiers

	Subcontractors		
Prime contractors	Subsystems	Components	Suppliers
Abrams tank Bradley fighting vehicle	Electro-optical systems Gas turbine engine Transmission Radio Navigation unit	Optical lens and mirrors Gun mounts Cannon Roadwheels Aluminum castings Turret ring casting Thermal imager & laser range finder[a] Displays	Hardware Aluminum Steel plate Machine tools Depleted uranium

[a] There are over 100 suppliers for the thermal imager and laser range finder in the Abrams tank alone.

SOURCE: Office of Technology Assessment, 1992.

oping and constructing armored vehicles, they could be replaced should the need arise, although at potentially high startup costs.

Similarly, reduced production will affect some subcontractors and suppliers more adversely than others. For example, the electronics and optics manufacturers for the Abrams tank and Bradley fighting vehicle support a number of other weapon systems and should be able to maintain at least some of their capabilities if these other programs are not cut excessively. But other subcontractors and suppliers might be forced out of business should the production of the Abrams and the Bradley be reduced too far. The failure of these firms would have serious consequences for the production of weapon systems. Like the prime contractors, however, many of these subcontractors and suppliers could be replaced by others in related lines of work, especially if the government is willing to buy from foreign manufacturers.[16] In the case of truly unique manufacturers, the government will need to take some action, such as subsidies, stockpiling, transfer of technology and government-owned equipment, or redesign.[17] As the production base shrinks, policymakers will face this issue again and again, in sector after sector.

Shipbuilding

The national shipbuilding industry is currently in a severe—some say terminal—slump. The bottom fell out of the commercial shipbuilding market in the 1980s. At the beginning of the decade, 69 commercial ships were either on order or under construction. By 1988 this number had fallen to zero. The order book remained blank until a single new ship was ordered in 1990.[18] If it were not for the U.S. Navy's pursuit of a 600-ship Navy, the U.S. shipbuilding industry might have completely collapsed from the lack of commercial work. (See figure 4-4.) Now, some analysts are projecting a reduction in naval forces to 400 ships or fewer, which will result in a further consolidation of the industry.

In shipbuilding, as in many other areas of defense contracting, there are significant differences in the structure and focus of an organization responding to the defense market as opposed to the commercial marketplace. Not only are there obvious differences in naval and commercial ships related to the installation of complex modern weapons, but the hull structure and machinery of warships are built to much more demanding specifications to provide resilience against blast damage, flooding, fire, and other hazards of combat. These differences demand a larger and more technologically advanced workforce at yards doing naval work.

16 For example, the aluminum roadwheels on the Abrams tanks are finished by Urdan Industries in Israel.

17 The President is authorized by P.L. 85-804 to grant extraordinary contractual relief to failing firms judged "essential to the national defense." This law was recently applied in the case of the Action Manufacturing Co. The slowing defense economy of the late 1980s and increased competition due to new laws requiring increased competition (CICA, to be discussed below) eventually forced this company to cease operations in 1989. This loss to the production base threatened to shut down or interrupt manufacturing at five Army ammunition plants and arsenals, and two contractors. Action was awarded relief on the grounds that the company was essential to the national defense because of its impact on mobilization, other producers, and readiness. (U.S. Congress, General Accounting Office, "Army Contract Adjustment Board: Decision to Grant Contract Relief to Action Manufacturing Company," GAO/NSIAD-91-230, July 1991, pp. 1-2 and 8-11.)

18 U.S. Navy, Naval Sea Systems Command, Corporate Operations Directorate, "U.S. Shipbuilding Industrial Base, 1980-1990," briefing book, July 1990.

Figure 4-4—Navy and Commercial Ships Under Construction, 1980-90

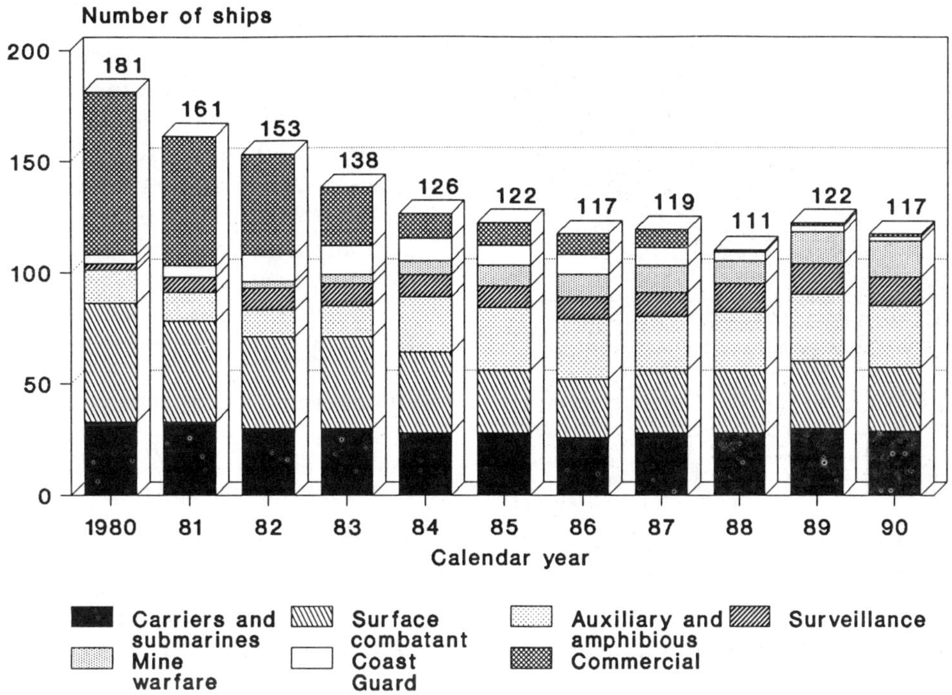

SOURCE: Naval Sea Systems Command, July 1990.

Table 4-2—Endangered Navy Shipbuilding Support Industries

	Domestic manufacturers			
Products	1980	1985	1990	2000
Boilers	3	3	2	1
Air circuit breakers	1	1	1	1
Condensers	8	8	6	4
Large diesel engines	3	2	1	0
Periscopes	2	2	2	2
Propellers	9	9	7	5
Reduction gears	9	9	9	2
Large shafting	6	5	4	3
Steam turbines	3	3	3	2
Power distribution switchboards	11	10	9	6

SOURCE: Naval Sea Systems Command, July 1990.

Shipyard subcontractors and suppliers, which can number in the thousands for a complex naval vessel, also face a difficult future. The number of primary subcontractors is expected to fall by the year 2000 to less than 75 percent of 1990 levels.[19] Moreover, many critical subtier vendors are dependent on a single class of vessel for their continued existence. Table 4-2 lists some of the more threatened capabilities, most of which are older technologies. Analysts are also concerned about the future of companies involved in nuclear ship propulsion. The Bush Administration's revised 5-year shipbuilding plan for fiscal years 1993 to 1997 includes only one nuclear-powered ship, an aircraft carrier, and no nuclear-powered submarines. The supporting nuclear propulsion companies have no civilian market to fall back on in a period of decreased shipbuilding. Nuclear-qualified shipyards may find some work in overhauls and decommissioning nuclear-powered vessels being taken out of the active fleet.

[19] Ibid.

Photo credit: The DoD

The Aegis guided-missile cruiser *Antietam* (left) prepares to be launched from the Ingalls shipyard, while work continues on the *Leyte Gulf* (right).

As with the other sectors discussed, the government can adopt measures to maintain domestic capabilities or, in some cases, fill gaps in production through foreign sourcing. For example, the U.S. Navy already depends on foreign purchases of large diesel power plants, periscope lens glass, and large seamless pipe.[20] Subcontractors and suppliers in advanced technology will, however, have substantial markets abroad.

Some analysts project a resurgence of the international shipbuilding market in the second half of the 1990s because of the growing obsolescence of the world merchant fleet. For the United States to take advantage of this trend, commercial yards will have to survive until the upturn begins and then build ships that are cost-competitive both in unit price and financing arrangements. Competition will be tough, particularly in the construction of technologically unsophisticated ships (e.g., single-hull tankers). U.S. shipbuilders might be advised to concentrate on those ship types that require more expertise (e.g., double-hull tankers, refrigerator ships, liquid-chemical container ships, and self-unloaders). They could also profit from increased foreign sales or subcontract work on naval vessels. If any of this is to occur, however, industry analysts argue that some governmental action will be required to break down foreign shipbuilding subsidies, primarily in Western Europe, and to promote foreign military sales.[21]

Conventional Munitions

The conventional munitions sector differs from the other industrial sectors cited above in that it produces in large quantities, often in the millions per year. In 1985 the U.S. ammunition budget was about $5 billion; in 1991 it had fallen to $2.3 billion with further reduction expected.[22] As in other sectors, there is considerable overcapacity in munitions production, including some mothballed plants. The military requirement for munitions is the sum of peacetime replacement needs (from training and testing) and war reserve requirements. This requirement has generally not been fully funded in the past and, with ever tighter procurement budgets, is unlikely to be fully funded in the future.[23] Moreover, surge efforts in preparation for the Persian Gulf War filled inventories with ammunition that went largely unused. Angelo Catani, President of Olin Ordnance, described this situation as ''accelerating our way out of the business.''[24] The Persian Gulf War also validated ''smart munitions'' at the expense of traditional ''dumb munitions.'' These smart munitions are produced in smaller quantities and have higher unit costs, with most of the cost going for guidance systems and not explosives.

Olin Ordnance, as one of the three domestic manufacturers of medium- and large-caliber ammunition, plans to survive the changes in defense production by restructuring and downsizing, and exploring new markets, such as ordnance disposal and environmental cleanup.[25] However, the oppor-

[20] Ibid.

[21] Shipbuilders Council of America, ''Update on World Shipbuilding Subsidies,'' special report, March 1991.

[22] ''From the Boardroom: Angelo A. Catani, President, Olin Ordnance,'' *Armed Forces Journal International,* October 1991, p. 73.

[23] Kenneth Girardini, ''The Army's Conventional Munitions Acquisition Process,'' Rand Note N-2864-P&L, The Rand Corp., July 1989, p. v.

[24] ''One on One,'' an interview of Angelo Catani, *Defense News,* July 1, 1991, p. 23.

[25] ''From the Boardroom,'' op. cit., footnote 22, p. 74. As this report went to publication, Alliant Techsystems, another domestic ammunition producer, had tentatively agreed to purchase the defense operations of Olin Corp.

Photo credit: DoD

The three man crew of the Multiple Launch Rocket System can fire 12 rockets over 30 km. in less than a minute.

tunities for saving the production facilities themselves are severely limited by the specialized nature of the manufacturing equipment and the lack of commercial markets for large ammunition.

When Preserving the Current Base Is Not Critical

Maintaining all parts of the current production base for a given weapon system might not always be necessary, particularly if the Administration and Congress adopt a long-term, mission-oriented approach to defense procurement. In many areas, reduced U.S. forces can be equipped from the current stockpile of weaponry for years. The technological edge of these systems could be assured through periodic upgrades, as described in chapters 3 and 5.

Meanwhile, design, development, manufacturing, and maintenance engineers could build and test experimental weapon prototypes that emphasize affordability, producibility, usability, and maintainability, in addition to performance (see ch. 3 for a discussion of prototyping). These prototypes may be direct descendants of current systems (e.g., a prototype follow-on to the Abrams tank) or they may achieve the mission of the current system in a new way (e.g., the Multiple Launch Rocket System v. traditional artillery).

Photo credit: TRH, London

The soldier of the future will live in an world vastly different from our own.

When a prototype has been sufficiently tested and a requirement appears, the new weapon system could enter production and replace aging equipment. **Since the new system could be truly revolutionary in design, its production base might be substantially different from the current base.**[26] (For example, the ability to cast large steel turret rings would not be needed to manufacture a ceramic, turretless tank.) New systems could be designed with common components, thereby simplifying and concentrating the production base and making lower production rates economical. Many firms working on current systems will recognize these shifts in

[26] In the past, few new weapon systems have been so revolutionary that the old production base was bypassed entirely. Products have tended to be more evolutionary as producers of old components have moved on to new components. This will probably be true in the future as well, although a series of prototypes may advance to the point where their basic hardware differs from the last produced model. Moreover, the new system may have little in common physically with systems preceding it (e.g., atomic weapons and guided missiles).

Box 4-A—Competing Goals of Defense Production

Efficiency is not the only grounds on which to judge the defense production base. Other public interest goals have been important factors in the structure of the current production base. These goals include:

- Maintaining employment levels and geographic distribution.
- Providing workers with education and skills.
- Fiscal accountability and safeguarding the taxpayers' money.
- Supporting small and disadvantaged businesses.
- Stimulating the national economy.
- Competition to ensure fairness and access.
- Buying American products to protect American jobs.

In a period of much reduced defense spending, policymakers might choose to adopt efficiency as the prime goal for the future defense production base to ensure that limited defense funding provides the maximum defense capability. Political realities, however, make it unlikely that the influence of public interest goals on defense procurement will disappear completely. As defense resources become increasingly stretched, policymakers may choose to elevate the relative importance of efficiency in restructuring the base. At a minimum, it might be necessary to make efficiency paramount in critical sectors where the future production base is especially threatened. Congress might review DoD efforts to identify vulnerable portions of the base where alternative sources are not readily available or are politically unacceptable (e.g., sole foreign suppliers) and request further studies if these efforts are found deficient. Then, Congress might exempt critical firms or the defense industry as a whole from public interest laws and regulations.

This report leaves the judgment of the appropriateness of various public interest goals to policymakers and focuses solely on options for producing "the most bang for the buck."

SOURCE: Office of Technology Assessment, 1992.

production opportunities and redirect themselves of their own accord.

EFFICIENT PRODUCTION

An efficient peacetime production base is defined as one that manufactures materiel that is affordable, manufacturable, usable, maintainable, and of good quality. (Two other desirable characteristics of the future production base—responsiveness and mobilizability in crisis or war—are addressed in subsequent sections). Congress, the Administration, and private industry can adopt several measures—separately or in combination—to create an efficient production base for the 21st century. The measures include: streamlining production and consolidating industries; operating at lower production rates; shifting away from the manufacture of end items toward prototypes, upgrades, spare parts, and maintenance; reducing barriers to civil-military integration; cooperating with allies; stimulating innovation; and increasing procurement and equipment commonality. The goal of efficient production is sometimes in conflict with other interests that have shaped the production base in important ways in the past. This conflict is discussed in box 4-A.

Streamline and Consolidate Industry

Streamlining and consolidating the current base are essential for efficient production. Defense manufacturers across the board are streamlining their operations. They are trying to sell off excess facilities, laying off or retiring workers, and diversifying into other businesses. Some companies have abandoned defense work; resulting in a consolidation of their industrial sector.[27]

Attrition will eventually reduce the size of the production base to a level consistent with decreased defense spending, but a lack of long-term planning will leave the base weaker and potentially crippled in key sectors if important manufacturers fail. The DoD did not regard this as a major concern until recently. In a report to Congress last November, the DoD wrote: "In a broad context,

[27] The Industry, Technology, and Employment Program at OTA is engaged in a companion study on how to smooth the transition of businesses and personnel into the private sector. Their first report is U.S. Congress, Office of Technology Assessment *After the Cold War: Living With Lower Defense Spending*, OTA- ITE-524 (Washington, DC: U.S. Government Printing Office, February 1992).

[28] "Report to Congress on the Defense Industrial Base," op. cit., footnote 15, p. ES-7.

free market forces will guide the industrial base of tomorrow.''[28] The DoD argued that active government intervention in the defense market would only be required in areas where technological or manufacturing capabilities critical for national security were threatened. The guiding principle at the DoD had been that the government is not ''wise enough'' to pick winners and losers and that, for the most part, market forces should make these determinations.[29] In recent months, however, DoD officials have begun to discuss more active options for preserving portions of the DTIB, including a prototyping strategy of sorts.[30] However, outside analysts argue that the coming budget reductions will be larger than the Bush administration is planning for and that unless decisive action is taken to protect the future base, it will be severely undermined.

Internal Streamlining

The government can influence the internal streamlining of individual firms in several ways. It can stabilize the business environment by making more reliable force projections and by predictable multiyear program funding. It can reduce administrative barriers that block the integration of commercial and military production. And it can support the transfer of relevant manufacturing technology. (Each of these actions is discussed below.) Most internal streamlining, however, must be company-initiated to enable firms to compete for fewer and smaller defense contracts.

Consolidation of Industrial Sectors

The government can have a more direct impact on the degree of consolidation of defense industrial sectors. In industries where future procurement will be much smaller than present production levels, the government might decide to pursue policies that ensure only that the best manufacturers survive, even if this means that others do not. For example, the fighter aircraft industry now consists of seven prime contractors.[31] Reductions in the number of U.S. Air Force and Navy fighter wings in the future will probably force one or more of these companies to leave this business.

Defense firms are unlikely to leave the defense business readily (as box 4-B suggests). Government action might either prop up these companies (e.g., by distributing contracts to maintain their survival or by waiving competition requirements)—perhaps weakening all of them—or help encourage consolidation among the firms to a number more commensurate with demand.[32] Facilitating mergers among sector participants would foster consolidation and avoid the loss of unique capabilities and talents.[33]

Changing Competition Rules

Government rules and regulations could also be changed to emphasize maintaining the health of innovative manufacturers in critical sectors.[34] (Less critical sectors might also benefit from these changes, but the need for them might be outweighed by social considerations, as was discussed in box 4-A). *Redesigning Defense* reported that industrialists pointed to the Competition in Contracting Act (CICA) of 1984 (Public Law 98-369) as a major source of problems in the DTIB.[35] In their view, the focus of CICA on full and open competition based on low-price bidding instead of quality or past performance has allowed some unqualified and inexperienced companies to get into the defense business at the expense of established and reputable producers. Without the discipline of past performance evaluation, new competitors may submit bids

[29] ''Atwood: Purely 'Industrial Base' Contracts on the Horizon,'' *Aerospace Daily*, June 3, 1991, p. 370.

[30] ''Statement of the Secretary of Defense Dick Cheney,'' op. cit., footnote 13.

[31] Boeing, General Dynamics, Grumman, Lockheed, McDonnell Douglas, Northrop, and Rockwell.

[32] The Army Munitions and Chemical Command is using existing regulations and statutes, including the 1861 Arsenal Act, to consolidate its ammunition mobilization base.

[33] The geographical distribution of the remaining manufacturers is an important factor in consolidation planning. Concentrating them in one area allows workers to flow from one company to another according to production schedules.

[34] A DoD Advisory Panel on Streamlining and Codifying Acquisition Laws, composed of senior government and industry representatives, has been established by Congress to review all acquisition laws and offer recommendations for change where appropriate. This panel is subdivided into 6 working groups covering socioeconomic, contract formation, contract administration, intellectual property, standards of conduct, and other acquisition statutes. Final recommendations are expected in January 1993. (56 *Federal Register* 215, pp. 56635-56637.) This panel will not address how the DoD implements legislation. For example, single-source contracts are legal under specified conditions, however, procurement officers avoid such exemptions from usual practice because they raise the possibility of legal challenges by other producers.

[35] For a more detailed discussion of CICA see ''Box 4- B—Problems with the Competition in Contracting Act,'' *Redesigning Defense*, op. cit., footnote 1, p. 70.

326-447 - 92 - 4 : QL 3

Box 4-B—Teaming Arrangements

A few of the larger defense prime contractors have sought to avoid betting their future on all-or-nothing contracts by joining forces in teaming arrangements where several companies share the risk and rewards of competing for large contracts. A prominent example of industrial teaming occurred in the Advanced Tactical Fighter (ATF) competition, where virtually the entire fighter aircraft industry signed up to support one or the other (or both) ATF prototypes. The two teams together are estimated to have invested between $1.2 and $2 billion in the competition, with the winning team getting the chance to build perhaps the only new fighter this century. In late 1991, the Navy ran a similar competition for the AX attack plane and awarded concept exploration and definition contracts to the following teams of prime contractors (shaded) and major subcontractors:

Major Partners of AX Team Proposals

Team A	Grumman	Boeing	Lockheed		
Team B	Boeing	Lockheed	General Dynamics		
Team C	General Dynamics	McDonnell Douglas	Northrop		
Team D	McDonnell Douglas	LTV Corp.*			
Team E	Lockheed	Rockwell			

*As this report went to press, the ownership of LTV Corporation's Aerospace Division was being determined by bankruptcy proceedings.

SOURCE: *Washington Post*, 1991; *Defense News*, 1992; U.S. Navy, 1992.

Teaming arrangements can be good for industry and the DoD if companies with complementary skills work together to produce a system that no single company could build alone. Indeed, almost all major weapons are built by teams. In the AX competition, for example, Grumman, McDonnell Douglas, and LTV have past experience with the special demands of naval aviation; Lockheed and LTV have expertise in stealth technology; and Boeing is strong in avionics.[1] But teams may also be founded less on unique qualifications than on a desire to carve out a piece of a diminishing market and to share financial risks.[2] In such a case, the combined resources of the partners are less of an advantage and more of a burden: the extended collection of prime and subcontractors serves as a source of increased bureaucratic overhead, conflict between corporate cultures, and miscommunication (see figure below). This is particularly true in competitions like the AX where a single firm may compete on more than one team, requiring internal barriers to the transfer of information. Moreover, there is a possibility that partners competitive in other programs may withhold their best ideas and personnel from the team. In the end, however, no amount of teaming to win a piece of what will be a smaller contract pie will support the current-sized base. Government policymakers, understanding this, should be wary of awarding contracts to teams that do not have complementary technological strengths.

Interlinking Teams

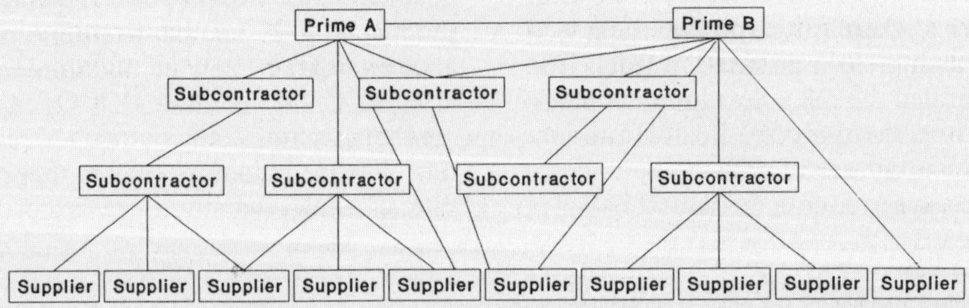

SOURCE: Office of Technology Assessment, 1992.

---

[1] Steven Pearlstein, "Strange Bedfellows Emerge to Chase A-X Bid," *The Washington Post*, July 21, 1991, p. H1 and H12; and Anthony L. Velocci Jr., "AX Competition Critical to Many Team Members," *Aviation Week and Space Technology*, July 22, 1991, pp. 18-19.

[2] In the ATF competition, each major player had a 50-50 chance of winning a share of the contract, instead of a 1-in-7 chance if the aircraft primes had gone it alone.

SOURCE: Office of Technology Assessment, 1992.

that they cannot fulfill without more funds later or transferring the contract to another manufacturer—all at added cost to the DoD.

Low-price bidding often favors companies that "build-to-print" (i.e., they produce off someone else's drawings), because these companies do not carry the high overhead cost of maintaining an R&D capability.[36] Build-to-print companies are an important segment of the current defense production base, often producing products of highest quality. However, **as industrial sectors consolidate, policymakers will need to decide whether it is better to support firms that develop new systems rather than build-to-print companies.**

One option for placing defense bidding on firmer ground lies in further efforts to change the criterion for contract award from lowest bid to "best value," based on the past record of a company in meeting price, schedule, and quality goals, the capability to do the job, and, in a few sectors, the importance of an individual company to the maintenance of the base.[37] New competitors could still be invited to submit bids, but they would need to be particularly innovative to overcome the handicap of no past performance record. Another option is the preselection of qualified bidders. The U.S. Navy applies a combination of best value and preselection in certain procurements. Rather than bidding for each contract as it comes due, firms are interviewed once and put on a list in order of their assessed ability to complete a job successfully. If the first company on the list refuses the contract, the next company is called in for negotiations. The list is then reused for similar procurements.

The U.S. Army is currently experimenting with alternative approaches to contract awards. One initiative has separated the risky development phase of procurement from the more predictable production phase. The uncertainties of development, combined with the pressure to submit the lowest bids, led to repeated cost overruns in the development phase in all Services during the 1980s. Cost overruns often hurt both the Services and the contractors, especially if the latter were working under a fixed-price arrangement with cost overruns charged against firms. The Army covered all development costs on its new light helicopter, the RAH-66 Comanche, and then looked for production bids that were realistic and demonstrated an understanding of the program. According to the Army, fully funding development reduces the likelihood of costly surprises and delays in production.[38] Although a team comprised of Boeing and Sikorsky was awarded the contract, recent budget cutbacks have left the future of the Comanche in doubt.

Public and Private Arsenals

In some cases, procurement might be so low that, even after extensive restructuring, there is only enough work for one manufacturer. The result would be what is known as a "natural monopoly."[39] For easily produced items, a natural monopoly would still allow competition, since other companies could bid to take over the contract when it expired. But for items requiring special machinery and skills, competition would have to be induced artificially and the government would have to absorb the costs of helping a new contractor develop the capability to produce these items whenever the contract changed hands. Moreover, the changeover might leave gaps in production capability. As the dominant buyer of defense equipment, the government would retain significant leverage for modifying the behavior and prices of a monopolist vendor.

In *Redesigning Defense,* OTA suggested that **private or public arsenals might be established in sectors where natural monopolies exist.** Private arsenals could be either GOCOs or companies that receive noncompeted contracts. Examples of important defense industrial sectors that might be forced into arsenal production are armored vehicles and

[36] If the policymakers took a radical approach to restructuring the DTIB and located all development responsibilities in R&D centers, then the distinction between build-to-print houses and other manufacturers would disappear (see ch. 3). Quality of production, however, would remain as a legitimate standard for consolidating the production base.

[37] The commercial strength of a company might also be an advantage in a best value competition. A diversified company might be better able to weather lulls in defense production than a less diversified company and maintain its defense capabilities without DoD support. In any case, best value criteria would need to be made explicit to limit subjectivity and deter legal challenges.

[38] Caleb Baker, "Army Calls for Changes in Contract Process," *Defense News*, vol. 6, No. 11, Mar. 18, 1991, pp. 4, 44.

[39] A natural monopoly "is a monopoly that occurs because it is economically impractical to have competition, as when the position of consumers would not be improved by having 30 water companies offering their services to every household in a certain city. The extra cost of installing 30 sets of pipe would more than offset any possible price reduction brought about by competition, so in the U.S. most natural monopolies are *regulated* monopolies." In Donald W. Moffat, *Economics Dictionary* (New York, NY: American Elsevier Publishing Co., 1976), p. 198.

ships. Designating one source for armored vehicles (e.g., tanks, personnel carriers, and self-propelled artillery) and funding it specifically to maintain its capabilities might be necessary to preserve manufacturing skills and guarantee a mobilization capability. In shipbuilding, rather than let the shipyards go out of business one by one, it might be in the national interest to select the most modern and efficient yards, or those with unique capabilities, as shipbuilding arsenals. These arsenals would likely be privately operated, possibly GOCOs.[40] The Navy is reportedly planning to streamline shipyards on both coasts and create administrative hubs that will reduce overhead at individual shipyards.

Closing or Mothballing Facilities

In instances where the current supply of a weapon system is sufficient for the foreseeable future, the government might shut down all current production facilities—perhaps mothballing them—and accept the substantial costs and delays of reestablishing production if the need ever arose.[41] The existence of an experienced workforce to revive a facility would depend on related work being conducted elsewhere.

Second-Source Contracts

The past emphasis on second-source contracts must also be reexamined in light of a perceived need to strengthen quality manufacturers. Second sourcing was intended to protect the military from the unanticipated loss of a manufacturing capability and to reduce unit costs by injecting competition into the procurement process. But unless carefully handled, second sourcing may be unfair to the original producer, may lower incentives for firms to invest in R&D, and may not result in lower real costs.[42]

The original developer of an item carries overhead costs (e.g., R&D and design teams) that a second-source producer, particularly a build-to-print firm, may not have. Moreover, just when the original producer is lowering its cost through greater manufacturing experience and higher volumes,[43] the second source must build or maintain facilities, train personnel, and often repeat the mistakes already made by the primary source, perhaps resulting initially in lower quality. **If the policymakers' goal is to forge a stronger future production base, the government could support fully those companies that develop new systems, rather than weaken them by giving second-source work to inexperienced firms or those without R&D capability on the grounds of furthering competition.** If second sourcing is still deemed important, then the second-source field could be limited to those producers that have a development capability. Another option for the DoD is to fully fund development, making it profitable in its own right, or to separate production from R&D through the establishment of independent design houses. (See ch. 3.)

Technical Data Rights

DoD procurement officers often demand all technical data as part of a production contract in order to establish second sources. (See ch. 2.) Companies report they have to release technical data rights to win a contract even when providing it is not legally required. **Prime contractors are largely unaffected by the technical data rights issue, since their major task is systems integration, which is a difficult capability to transfer. In contrast, subcontractors see their proprietary information in products and processes, many of which have commercial applications, as the primary feature that distinguishes them from their competition.** Build-to-print defense firms can use this data to undercut R&D-intensive companies in defense contract competitions based strictly on

[40] Although the U.S. Navy in the past constructed its own ships in naval shipyards, it has not done so since the 1960s. Currently, the naval shipyards specialize in overhaul and repair, while all new construction is done in private yards. Because it would be difficult and expensive for the naval shipyards to relearn how to build ships, any shipyard arsenals would most likely be established at private shipyards currently engaged in shipbuilding.

[41] For example, General Dynamics estimates that if its tank facilities were completely shut down, it would take a minimum of 4 1/4 years after a reopening decision to produce the first tank and 5 years to achieve a rate of 60 tanks/month. The components that determine this delay, however, could be identified (i.e., gas turbine engines and depleted uranium armor) and stockpiled to cut down the time to first-unit production.

[42] The extent of direct cost savings gained through second sourcing is very difficult to measure. A Rand Corp. study of second sourcing noted, "in some cases (especially in the procurement of major systems) it may be actually less costly for the government to forgo competition and rely on a single supplier." This is primarily due to the need to facilitize the second source and transfer manufacturing procedures. In order to achieve true cost savings, the planned production quantity must be sufficient to allow both firms to achieve maximal productivity and offset the additional costs of the second source. (See J.L. Birkler, E. Dews, and J.P. Large, "Issues Associated With Second-Source Procurement Decisions," R-3996-RC, The Rand Corp., December 1990, pp. v, ix, and 26.)

[43] The cost of production for most items decreases over time as experience is acquired and past mistakes are avoided. These savings are found in shorter production cycle times, fewer manufacturing defects and design changes, and less waste. This decrease in cost can occur regardless of second sourcing.

lowest price and to gain advantages in commercial markets.

In the future, innovative firms will have a greater stake in holding on to technical data rights to help their commercial work. They may therefore refuse to bid for defense work if the DoD enforces technical data rights rules as it does today. Useful alternatives would be for the DoD to:

1. put the burden of proof on the government to demonstrate an explicit need for access to proprietary data;
2. limit the requirement for proprietary data to certain vital components;
3. keep collected data confidential until needed, perhaps in escrow, with government access contingent on specific conditions;
4. let businesses withhold this information for a period of time (similar to a patent) that will allow them to develop more advanced capabilities (a relatively short period in the electronics field);[44] or
5. compensate firms financially or with advantages in procurement for the full value of their proprietary data.

Strategic Partnerships

In the future, fostering a quality, integrated subcontractor and supplier base will be at least as important as supporting the best prime contractors. Yet consolidation among subtier firms is inhibited by many constraints. Although the primes are not always legally bound to compete their subcontracts, many do compete subcontractor awards in the belief that otherwise they will lose the contract. The primes argue that competing subcontracts is often expensive and sometimes results in poor quality work by the low bidders, causing the primes to be blamed for schedule delays and cost overruns.

Spokesman for the primes argue that the subcontractor and supplier tiers should be rationalized through strategic partnerships. This is currently occurring in the civil sector, where commercial enterprises are shedding their past practice of competing for lowest-priced components in favor of long-term relationships with subcontractors and suppliers of proven quality and dependability. Spokesmen for commercial industry argue that these partnerships result in lowest real cost (e.g., they avoid costly mistakes, redesigns, and the cost of reworking defective incoming parts). The Federal Government could ensure fair pricing through competitive bidding at the prime level and periodic audits.[45]

Reduce Production Rates

Policymakers should plan for future defense production that will be much lower than present levels. Funding cuts and equipment surpluses generated by force reductions will continue to lower production. This is not to suggest that all production will cease. **Even with the largest cuts now forecast, the military will continue to purchase tens of billions of dollars worth of equipment annually.** Future forces will still need to be outfitted; aging stocks will need to be replaced periodically.

One way of maintaining manufacturing capability in the future DTIB would be to set low production rates in lieu of traditional rapid rates.[46] The decline of a major Soviet-size conventional threat has made it less important to produce systems rapidly.[47] Low-rate production would allow the DoD, with a lower procurement budget, to maintain core manufacturing personnel, equipment, and facilities. These would serve as the base for fulfilling surge, mobilization, or increased peacetime requirements. (See box 4-C.)

Whether or not low-rate production increases unit costs depends to some degree on when and how the production decision is made. If a decision is made during the design phase of a product, then the design can be optimized for existing low-rate manufacturing equipment and processes. Manufacturing facili-

[44] In fact, businesses often delay handing over technical data until they have developed a more advanced capability.

[45] Commercial businesses that are now organized in strategic partnerships feel they can avoid price gouging by their restricted subcontractor and supplier base without the extensive oversight common in defense procurement by negotiating a long-term relationship in exchange for a reasonable price.

[46] Low-rate production in this report differs from the concept of "low-rate initial production" (LRIP). LRIP is often intended as a trial period during which the manufacturing processes and equipment are validated and final design changes are made before shifting to a higher rate. In this report, low-rate production remains constant and does not assume higher future rates. Thus, the best manufacturing processes, facilities, and equipment for LRIP may differ significantly from what is needed for low- rate production. See *Defense Manufacturing Management*, op. cit., footnote 4, ch. 11, pp. 12-13.

[47] In the commercial sector, the speed with which new products can be brought to market is becoming increasingly important. Low-rate production, designed to maintain a critical production capability over an extended period of time, necessarily contradicts this trend. Joseph T. Vesey, "Speed-to-Market Distinguishes the New Competitors," *Research-Technology Management*, vol. 34, No. 6, November- December 1991.

Box 4-C—Low-Rate Production is Not Always the Best Option

During the transition to a smaller, more efficient future production base, not all manufacturing skills, equipment, and facilities will need to be actively protected by the DoD. Many capabilities may be relatively common in the defense or broader commercial base. As long as workers and equipment can be assembled in a timely manner when production is needed, the DoD need not take extraordinary action to preserve a continuous capability in the defense production base. (Policies addressed in this chapter for fostering efficiency and preserving capabilities, however, may still be beneficial in these noncritical areas as a means of lowering procurement costs and raising quality.)

Stretching out production is particularly problematic for the wide variety of defense items that are procured in small quantities of less than 100 (e.g., special aircraft, ships, and satellites). In such cases, the indiscriminate adoption of low-rate production could result in unrealistic work schedules and therefore needlessly expensive costs. Instead, these products should be procured in economic batches as long as their associated manufacturing capabilities can survive between orders through related work or because the manufacturing processes involved are relatively simple. The use of economic production rates applies to subtier firms as well as to prime contractors. For example, even if it was decided to manufacture attack helicopters at a slow, steady rate to maintain worker familiarity with the process, many basic subcomponents and hardware could be procured up front to lower costs. (Of course, savings for bulk purchases would have to be balanced against storage costs and the requirement for earlier spending.)

SOURCE: Office of Technology Assessment, 1992.

ties and staffs can be sized appropriately, thus minimizing unit costs.[48] But transforming an active high-rate production facility to a lower rate will in general be difficult, less efficient, and expensive, and will require an arduous transition period when excess capacity and workers are reduced.

Predictable funding through multiyear procurements would enhance low-rate production. Such funding would facilitate long-range planning and lessen firms' fears of failure; thus it would permit more aggressive restructuring. This should in turn result in lower unit costs than would occur otherwise.[49] The disadvantage of multiyear procurement is that it reduces government budget flexibility, front-loads costs, carries significant penalties to the government for contract cancellation, and may make it more difficult to institute late design changes.[50] Adoption of multiyear procurement would require a consistency of defense planning and funding that is not evident today.

The key to the success of low-rate production is to establish an acceptable minimum production rate. This rate will allow the prime contractor to remain profitable and obtain all necessary subcomponents and supplies. The rate will depend on the size of operations, the flexibility of the factory, and the nature of other products produced. Rates will also be affected by the adoption of measures discussed later in this chapter.

Detailed information will not only be needed about the lowest sustainable production rate of the prime, but that of suppliers and subcontractors as well. In some cases, the lowest rate may be determined by the need to keep production lines open for a critical subassembly. Alternatively, rather than produce more of the final integrated system, the government might find it cheaper to subsidize the manufacturer of this subassembly, find another company willing to produce it or a redesigned replacement at lower rates, move production into a

[48] The capital investment in facilities and equipment for a new product can be a substantial proportion of the total program cost. A company that agreed to low-rate production for a new product might limit itself to one production line, because that is all it needs to manufacture the item. Moreover, the company might choose to rely on flexible and existing equipment, rather than investing in specialized equipment, to lower costs and leave open opportunities for altering production processes over the duration of the extended production run. Under traditional procurement, the same company would more likely run several production lines using specialized equipment in order to receive the quickest return on its investment. Cutbacks in procurement to this company after it had made this investment would result in higher unit costs.

[49] Conversely, unpredictable funding and production rates will raise unit costs, because of an inefficient use of manufacturing resources.

[50] Karen W. Tyson et al., "Acquiring Major Systems: Costs and Schedule Trends and Acquisition Initiative Effectiveness," IDA Paper P- 2201, Institute for Defense Analyses, March 1989, ch. 6, pp. 1-11.

Table 4-3—Impact of Low-Rate Production on the Abrams Tank's Subcontractor Base

	Number of tanks produced per month	
Component subcontractor	10	30
Electronics/Optics		
Sterling Heights, GDLS	Some problems	Acceptable risk
Cadillac Gage, MI	Some problems	Acceptable risk
Cadillac Gage, OH	Significant risk	Some problems
Texas Instruments	Some problems	Acceptable risk
Smith Industries	Some problems	Acceptable risk
Kollmorgan	Some problems	Acceptable risk
Precision Sensors	Some problems	Acceptable risk
GE	NA	NA
J-Tech Associates	Some problems	Acceptable risk
Hughes Aircraft	Some problems	Acceptable risk
Computing Devices	NA	NA
Kearrfott	NA	NA
Complex machining		
Scranton, PA, GDLS	Significant risk	Some problems
Detroit, MI, GDLS	a	a
Lima, OH, GDLS	Some problems	Some problems
Basic materials		
Atchison Casting	Some problems	Some problems
Lukens Steel	Some problems	Some problems
Idaho, U.S. DOE	b	b
Weapons		
RIA	Some problems	Acceptable risk
Watervliet Arsenal	Some problems	Acceptable risk
Propulsion		
Textron Lycoming	Some problems	Some problems
Allison	Some problems	Some problems
Stanley	NA	NA
Urdan	NA	NA
FMC	NA	NA

KEY: Companies that are least affected by lower production rates are listed as having "some problems." Companies listed as being at "significant risk" will be most negatively affected, with a potential for facility shutdown. NA means not available.

[a] M1 assembly line closed in 1991.

[b] Scheduled to begin closing in December 1992.

SOURCE: General Dynamics Land Systems (GDLS) Division, 1991.

government-owned arsenal, or stockpile all that will ever be needed in a one-time ''life-of-type'' buy.

A company might implement its low-rate production in different ways: spread evenly throughout the year, in odd-sized batches as orders come in, or all at once in a short period to allow a shift to other products for the remainder of the year. The sole mandatory requirement would be that critical capabilities and skills are maintained from one year to the next. OTA asked General Dynamics Land Systems to estimate the impact of low-rate production on the Abrams tank contractor base. The results are shown in table 4-3.

Firms that emphasize flexibility in manufacturing organization, processes, and equipment will be well positioned for a transition to low-rate production, as well as to the production of new products. (See box 4-D.) **Flexible manufacturing systems enable businesses to build several different products simultaneously on the same line (or at the same stall) and to shift from one project to another with a minimum of expense and effort.** In the extreme, each item on a flexible line might be unique. While flexible manufacturing can be capital intensive, requiring new flexible machines, this is not true for all products. For example, the most flexible and cost-effective manufacturing method for one-of-a-kind satellites might be to build them by hand.[51]

[51] Sandwich shops are often used as an example of a completely flexible assembly line with no automation at all—customers have a wide variety of sandwiches and ingredients to choose from.

Box 4-D—Restructuring for the Future: BMY-Combat Systems

One company that has already successfully navigated the transition to low-rate production is BMY-Combat Systems. In an interview in *Armed Forces Journal International,* the president of BMY revealed that the company reduced the production rate of the M-88 tank recovery vehicle from 20 per month to 3 to 4 per month and remained profitable. This rate reduction was one aspect of a company-wide strategic restructuring.[1] The success of this effort depended on a total restructuring of the production process so that six fairly similar products that use many common parts and processes could be produced on the same line. It also required the infusion of $80 million for plant modernization from BMY's parent corporation, Harsco; the replacement of most government tooling with the company's own, more flexible tooling;[2] a consolidation and rationalization of facilities; a 50 percent reduction in workforce; and a relative increase in foreign sales of its products from about 40 percent in 1985 to about 65 percent in 1991. The company can take small orders (5 to 6 vehicles) and integrate them with orders for other vehicles to maintain the production line. With the restructuring almost complete and a couple of years' orders on its books, BMY-Combat Systems' employment is now rising again. The firm's main concern about future production is how to ensure continued supply of key parts from subcontractors and suppliers. Commonality of products has allowed BMY to award multiyear contracts to its subtier vendors, but some of them are on the verge of going out of business.

[1] John G. Roos, "From the Boardroom: Barret W. Taussig, President BMY-Combat Systems," *Armed Forces Journal International,* May 1991, p. 27; and an OTA interview with BMY-Combat Systems officials on Nov. 14, 1991.

[2] Tooling that is purchased as part of a government contract remains government property and cannot be used for other work—governmental or commercial—without compensation. Companies will often purchase their own tooling and forgo government equipment to avoid the inflexibility of having tools that can only be used for one purpose without more paperwork and negotiations. This tooling is added to overhead charges.

SOURCE: Office of Technology Assessment, 1992.

Shift Business Focus

A company can lower its minimum viable production rate for a specific product by expanding its range of activity to include prototyping, upgrades, spare parts, and maintenance, and by manufacturing multiple products. Chapter 3 discussed the implementation of a prototyping-plus strategy that would provide certain manufacturers with the opportunity to build technology demonstrators or even an entire operational unit of prototypes prior to force modernization as a means of both fostering innovation and supplementing or temporarily replacing limited production.

In addition, a reexamination is in order of the DoD's practice of awarding spare-part production contracts to firms other than the original manufacturer. Although intended to increase the number of sources of supply and lower costs, this practice also has the effect of supporting build-to-print shops with little or no design capabilities, at the expense of the original manufacturer. In the commercial world, spare-part sales are often an important source of income to the original producer. Similarly, upgrade contracts that would have been competed in the past might be directed to a particular manufacturer, or competition might be limited to quality producers to help alleviate losses in production. Spare parts and upgrades can be a significant fraction of an industry's business. For example, the commercial market for spare parts for large aircraft engines is about half as large as the market for engines.[52]

Another important shift in business focus is for the DoD to transfer some depot-level maintenance work from the public sector to the private, as discussed in chapter 5. One reason to do this is to augment the dwindling workload of the original equipment manufacturer. As systems become more sophisticated, and perhaps modular, they could be returned to their originating factory for maintenance rather than duplicating this capability at government depots and shipyards.

Companies might also adapt to procurement shortfalls by diversifying their product lines. In some industrial sectors, market attrition or government policy might result in a consolidation of manufacturers such that products previously spread

[52] Pratt & Whitney predicts that the relative size of this market will grow well into the 21st century. Projections of the military market for large engine spare parts are clouded by procurement decisions, but they range from about 35-100 percent of the military engine market for the same period. Pratt & Whitney briefing, West Palm Beach, FL, Sept. 10, 1991.

Figure 4-5—Hypothetical Shift in a Defense Manufacturer's Activities

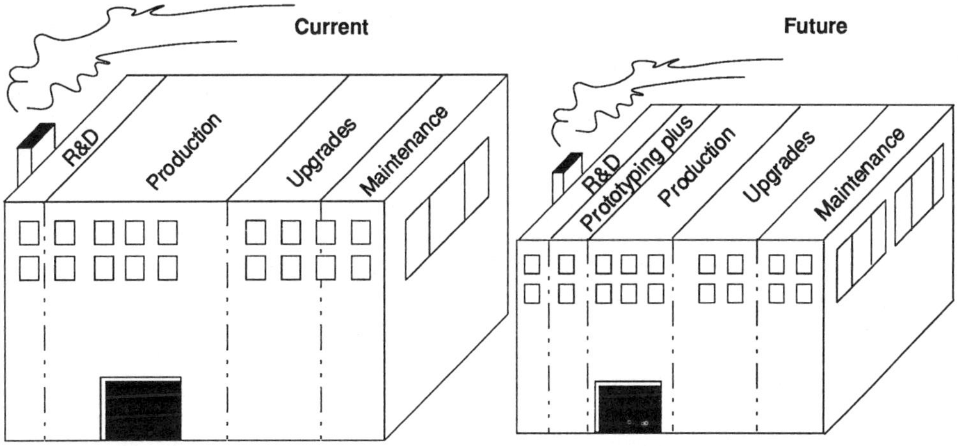

NOTE: Proportion of activities varies by individual sector and individual firm.
SOURCE: Office of Technology Assessment, 1992.

over several companies (e.g., ships and ammunition) become the domain of a few companies. A company might also seek to expand into other government or commercial activities (described below). Companies that are flexible in their manufacturing processes and that utilize similar equipment and skills will avoid many of the costs associated with starting new production.

Figure 4-5 depicts how a company might shift its business focus. Caution must be used, however, in any attempt to substitute other activities for production. For some industrial sectors (e.g., defense electronics) and types of companies (e.g., subcontractors) such activities are reported to be a viable way to survive lulls in production. Yet, for companies that produce complex integrated systems or products that require little maintenance, this option may not be a feasible way of preserving their full range of critical manufacturing skills (see box 4-E), facilities, and equipment. Companies whose manufacturing capabilities are critical to the base must be examined on a case-by-case basis to determine the viability of this strategy.

Reduce Barriers to Civil-Military Integration

The efficiency of the future defense production base may also be enhanced by changes in the relationship between industry and government. *Redesigning Defense* reported a broad consensus that government/industry relations have become increasingly adversarial.[53] This stems from laws and regulations adopted in response to public fears of waste, fraud, and abuse. Legislation, regulations, and the resulting procurement culture have led to voluminous contracts, layers of restrictive product specifications and auditing procedures, and barriers to communication among industrialists and with government program officers, and it has impeded off-the-shelf purchasing.[54]

Companies expend enormous energy and time on the paperwork associated with DoD contract bidding and auditing. These costs are included in overhead and ultimately added to the price of procured items. While large firms can "afford" specialized staffs to cope with this paperwork, smaller firms face a disproportionate burden. Paperwork requirements

[53] The Persian Gulf War eased relations temporarily. Procurements that usually take months or years were sped through in weeks. Spare parts and upgrades were rushed to the front in record time. Food, fuel, water, and other commodities were bought in local markets in Saudi Arabia and the United States until logistics officers could catch up with the rapid buildup. And commercial items were used to bridge gaps in Service procurement. For example, the Navy's Safety and Survivability Non-Development Item Office bought a number of products commercially, such as fire-fighting and detection equipment, air hammers, and body armor, and delivered them to the fleet within 45 days. (U.S. Department of Defense, *Conduct of the Persian Gulf Conflict: An Interim Report to Congress*, July 1991, ch. 8, pp. 1-4.)

[54] These issues will be discussed in more detail in ch. 6, but are presented here as they relate to production specifically.

Box 4-E—Worker Skills in the Defense Production Base

The streamlining of the defense production base has forced many companies to focus on short-term survival over long-term health. In addition to eliminating excess capital equipment and facilities, these companies feel compelled to reduce personnel costs. Reductions have focused largely on the early retirement of older, more experienced workers and layoffs of young new talent. There has also been a retrenchment in spending on worker training programs and apprenticeships. Many who have benefited from these services in the past are now moving into the commercial sector. Moreover, nondefense workers and students, seeing the downward slide of defense procurement, are looking for careers elsewhere.

In the short term, this situation is tolerable to the DoD, if difficult for some of the workers involved. The future production base will not need the number of people currently engaged in defense manufacturing. However, employee reductions that do not take into account future needs may undermine the long-term health of the base. Manufacturers need to retain their most qualified personnel, while production base planners need to ensure a continued supply of manufacturing talent. If the base becomes more commercialized, free market competition may be sufficient to generate the necessary talent. Government support may be needed to preserve select critical skills—from shop floor machinists to naval architects. This help could take the form of scholarships or trade school subsidies to employees or grants or tax breaks to businesses having trouble finding workers trained in needed skills. Alternatively, textual, audiovisual, and computer methods for storing manufacturing experience could be funded. Colleges and universities could strengthen the Nation's future production base by emphasizing manufacturing in engineering and business school curricula. A better educated workforce will make the future base more flexible.

SOURCE: Office of Technology Assessment, 1992.

have also deterred some commercial companies from bidding on defense contracts.[55]

Further inefficiencies result from the DoD's requirement that manufacturers produce items according to military specifications that dictate every facet of a product, including acceptable manufacturing processes. Many studies argue that these specifications are out-dated and overly rigid.

Many companies have segregated their defense and commercial work because of DoD requirements for specialized military manufacturing processes and the need to avoid burdening their commercial work with military accounting requirements. This segregation might entail separate production lines on the same shop floor, separate production facilities, or even totally distinct operating divisions within a company. **Segregation can create redundancies in equipment, personnel, facilities, and management, and create barriers to communication between military and civilian operations.** In the extreme, manufacturers in a defense division may have no direct contact with their counterparts in a commercial division. Such segregation raises costs and hinders the transfer of technology between the commercial and defense sectors.

Policymakers need to reassess the tradeoff between the costs of fraud and abuse and the cost of oversight to prevent them. While preventing abuse is an important task of government, the cost of current efforts to do so, both direct and indirect, are large and may outweigh the monetary and moral gains of catching the abusers. Program officers, auditors, and inspectors engaged in often uncoordinated and overlapping jobs pervade the defense industry. Congress might commission an in-depth study of the direct and indirect costs and benefits of military contract oversight. Accountability must be maintained, but policy might be redirected towards punishing transgressors more severely and rewarding responsible businesses, perhaps by making past behavior an important factor in awarding best-value contracts.

Government action to facilitate a more efficient integration of civil and defense industry can range from mild corrective measures to a radical restructuring of defense production. Some sectors of the defense production base are more amenable to integration than others (e.g., defense electronics v. shipbuilding), making sweeping decisions more difficult.

[55] For example, Lotus Corporation spent over $2 million trying to comply with DoD accounting system standards before giving up. Hewlett Packard only deals with the DoD on a commercial basis. See Jacques S. Gansler, "Restructuring the Defense Industrial Base," *Issues in Science and Technology*, Spring 1992, p. 51.

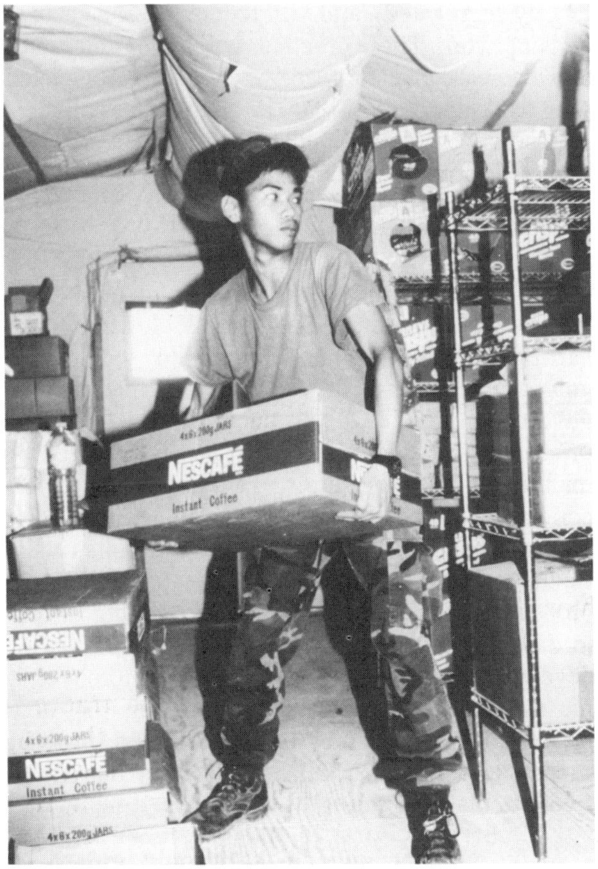

Photo credit: The DoD

Soldier stacks provisions in a food tent in Saudi Arabia.

The DoD has not ignored the complaints of industry and has made repeated efforts to smooth relations. In recent years, it has adopted a few programs designed to lessen the intrusiveness of government oversight. The Qualified Manufacturing Line program for the semiconductor industry, which began in 1987, is one example. Under this program, a company can demonstrate that its production lines meet military standards and thus avoid having to test each chip, as has been traditional. The electronics industry has created the National Electronic Component Quality Assurance System as a self-policing measure. Among other activities, this system conducts an audit of the supplier base that used to be done by each company individually.[56] Neither of these programs is widespread at this time.

Industry has also joined with the defense acquisition, inspectorate, and auditing communities in the voluntary Contractor Risk Assessment Guide (CRAG) program to reduce oversight burdens on businesses. The CRAG outlines five critical auditing areas: indirect cost submissions, labor charging, material management accounting systems, estimating systems, and purchasing.[57] If a company demonstrates effective internal accounting controls in one or more areas, the Defense Contract Audit Agency (DCAA) reduces its oversight of that area. While the CRAG program has been implemented successfully in a few tens of companies, the incentives for businesses to participate are mixed. On the positive side, the CRAG program potentially reduces the DCAA presence at a company; on the negative side, companies may find themselves paying for internal auditing services that the DCAA would otherwise do for free, while the presence of oversight officials from other government agencies is not reduced.

The defense production base could also take greater advantage of products already on the commercial market. Efforts to increase the percentage of DoD products bought from the private sector have been under way for almost 20 years.[58] DoD Directive 5000.1 requires that the "maximum practicable use shall be made of commercial and other nondevelopmental items." For example, the Navy is installing commercial computer systems on combat ships for many functions that were formally performed by unique, Navy-designed computers.[59] Off-the-shelf procurements can be facilitated in a number of ways, including: 1) elimination of unreasonable specifications that block commercial purchases (e.g., requiring nuclear effects hardening on

[56] See Debra Polsky, "DoD Chip Oversight Plan Gains Favor," *Defense News*, July 1, 1991, p. 9.

[57] Defense Contract Audit Agency, "A Report on Activities," DCAAP 7641.81, March 1990, p. 3, and Department of Defense, "The Contractor Risk Assessment Guide," October 1988.

[58] U.S. Congress, U.S. General Accounting Office, "Defense Acquisition Reform: Implementing Defense Management Review Initiatives," GAO/NSIAD-91-269 (Gaithersburg, MD: U.S. General Accounting Office, August 1991), p. 11.

[59] Neil Munro, "Navy Wants More Commercial Computers Aboard Ships," *Defense News*, Jan. 27, 1992, p. 30.

products that are a part of an unhardened system),[60] and 2) redesign of systems to use off-the-shelf components (e.g., instead of making a monitor rugged, enclose a commercial TV in a shock-resistant shell). Buying many copies of a commercial product instead of one militarily unique product might save money and provide sufficient redundancy to overcome a lack of ruggedness or capability.[61] Of course, many items will still have to be procured according to military specifications to ensure that they will perform reliably under demanding field conditions.

The efficiency of the peacetime defense production base may be greater if the relationship between government and defense manufacturers moves to a more commercial basis. Manufacturers would be evaluated on their ability to produce items on time, at an agreed price, and of agreed quality. Military specifications would focus more on the essential characteristics of form, fit, and function, and less on laying out explicit manufacturing procedures. For many products, commercial seals of approval or certification might be sufficient (e.g., Underwriters Laboratory or the International Organization for Standardarization). It might be important for the DoD to continue outlining some critical procedures (e.g., specialized welding), but these specifications could be arrived at through negotiations with the manufacturer, who might know of superior processes. Military specifications should only be passed on from one project to another if they are demonstrably relevant.[62]

Several studies argue that increased reliance on commercial business practices will give industry more flexibility in carrying out contractual obligations, will allow integration of commercial and defense facilities, equipment, and supplies (now largely segregated), and will reduce overhead spent on paperwork. These practices will lower unit prices and decrease the competitive penalties associated with working for the DoD. The adoption of commercial practices could open competition to a wider circle of companies, because nondefense firms that presently avoid defense work would take advantage of the new environment.

For defense companies with little or no experience with commercial operations, lowering the barriers to civil-military integration will create unique challenges. In the defense market, the government defines the product, determines appropriate pricing and profit margins, and specifies manufacturing procedures. The challenge to the company is to convince the government that it can develop the product more efficiently than its rivals. In the commercial market, on the other hand, a producer must rely on its own resources and insight to define products and must carry the full cost of developing them until they can be sold. Moreover, it must convince customers to purchase the new product and possibly provide warranties and product support as a consumer relations strategy more than a contractual obligation.[63]

At the extreme, policymakers could undertake a radical restructuring of the DTIB based on the wholesale elimination of administrative and legal barriers to civil integration, combined with the redesigning of military systems and the acceptance of form, fit, and function specifications. These policies would effectively merge the DTIB into the national industrial base. Many defense items could be close variants of commercial products (e.g., the KC-10 tanker is based on the DC-10).[64] Products that are truly militarily unique would still be built, but could take advantage of

[60] Some commercial products, notably in the aerospace sector, are manufactured at or above military specification standards. The DoD might be able to safely buy these items off-the-shelf at a lower cost. Periodic testing or performance reviews would ensure quality control. In other areas, a lowering of military requirements might make the product sufficiently inexpensive to procure enough copies of an item to compensate for the lowered specifications. Representatives of the shipbuilding industry have suggested that specifications on sealift vessels be lowered in this way so that more such vessels can be procured. These vessels would be protected by combatant ships built to military specifications.

[61] During the Persian Gulf War, the DoD bought 10,000 commercial light-weight, global positioning system receivers to help troops, particularly in helicopters and tanks, locate themselves in the featureless open desert. See *Conduct of the Persian Gulf Conflict,* op. cit., footnote 53, ch. 8, p. 3.

[62] One proposal being discussed in the Pentagon calls for a "zero basing" of military specifications. This proposal would reverse the incentive structure of the current acquisition system by forcing program officers to defend the application of any military specification to a product. See Lucy Reilly, "Milspecs Go Under the Gun: Pentagon Considers 'Zero-Basing' Approach," *Washington Technology,* vol. 7, No. 2, Apr. 23, 1992, pp. 1, 13.

[63] Defense manufacturers are increasingly being asked to provide warranties for their products. If these warranties are paid for in a production contract, they may be one way to shift maintenance work back to the original manufacturer.

[64] McDonnell Douglas is developing a new helicopter, the MD-900 Explorer (formerly the MDX), justified on commercial grounds alone, but said to be sturdy enough for utility, armed scout, medical evacuation, and other military missions. See Frank Colucci, "MDX in Uniform," *Defence Helicopter,* vol. 11, No. 1, March-April 1992, pp. 24-27.

Photo credit: U.S. Air Force

McDonnell Douglas KC-10 Extender refuels F-15A Eagle fighter.

many of the same production facilities, equipment, and workers as commercial products.[65] Unique military specifications on products or processes would be an exception. Any activities that require secrecy could still be segregated, but only when absolutely necessary. Future flexible factories and other innovations described below would increase the likelihood that this radical strategy would succeed.

Cooperate With Foreign Nations

The U.S. defense production base can be bolstered through making greater use of the international base. First, sales of military equipment to foreign countries can be used to keep production lines open that might close otherwise.[66] Second, U.S. firms might increase cooperative activities with foreign countries, perhaps establishing joint ventures. Third, the DoD might purchase items overseas that either are not produced domestically or are cheaper abroad. Fourth, foreign business might purchase U.S. defense enterprises that are failing and make them productive again. All of these activities are currently under way and raise opportunities and concerns. **Policymakers will need to evaluate how changes in these activities might affect the production base and what their foreign policy and national security implications might be.**

As domestic procurement declines, foreign sales may be one way to keep production lines running. A few defense firms already produce a majority of their equipment for export. As one industry trade group official stated, "Exports are no longer just the icing on the cake. They are the cake."[67] For many companies, exports have become relatively more important as domestic sales have declined. For example, General Dynamics projects overseas sales to increase from 17 percent in the mid-1980s to about 50 percent in the mid-1990s, while Martin Marietta plans to move from 8 percent in foreign sales in 1991 to about 20 percent in 1994.[68] Firms were particularly optimistic about future sales after the success of U.S. armaments in the Persian Gulf War. Arms sales to foreign nations may also provide the United States some political leverage over recipients through the sale of upgrades and spare parts.

There are, however, two major problems with an expansion of foreign sales. First, as a result of the end of the cold war, many countries, particularly NATO and former Warsaw Pact nations, have less need for weapons.[69] The shrinking world market for weapons is increasing global competition. Second, the spread of advanced weapons technology around the globe has raised concerns over weapon proliferation and the threat modern U.S. weapons may pose to U.S. forces engaged in future conflicts.[70] Without the global Soviet threat, calls for regional bans on

[65] In the few instances where a commercial remedy could not be found, the government might establish a public or private arsenal (e.g., nuclear submarines).

[66] For example, Air Force Secretary Rice testified before a Senate Appropriations subcommittee that the General Dynamics F-16 fighter production facility in Fort Worth, TX will stay warm based solely on foreign sales and aircraft upgrades when the Air Force cancels F-16 orders after next year. The Air Force may need the plant again for a new F-16 variant sometime in the future. A General Dynamics official and some members of Congress were skeptical of this approach. See Ron Hutcheson, "Plan Threatens GD Plant, Official Says," *Fort Worth Star-Telegram,* Mar. 18, 1992, p. 1.

[67] Joel L. Johnson, international vice president of the Aerospace Industries Association, cited in Steven Pearlstein, "Struggling to Keep Weapons Programs Alive," *The Washington Post,* Mar. 9, 1991, p. C2.

[68] Ibid.

[69] Developing countries, which have been less affected by the political changes in Europe, accounted for over 75 percent of arms import sales in 1988. U.S. Bureau of the Census, *Statistical Abstract of the United States: 1991,* 111th ed. (Washington, DC: U.S. Government Printing Office, 1991), p. 339.

[70] For more information on the negative aspects of arms transfers, see U.S. Congress, Office of Technology Assessment, *Global Arms Trade,* OTA-ISC-460 (Washington, DC: U.S. Government Printing Office, June 1991).

weapon sales may inhibit the traditional desire to fortify friendly nations against their adversaries and block sales to regions that still demand new weapons.[71] Congress will have to weigh the importance of controls on international weapon exports against the risk that such controls will be circumvented and the need of the defense production base and U.S. allies for arms sales.

Moreover, the decline of the world arms market is consolidating defense industries on a global scale. In Europe, defense companies are increasingly engaging in transnational mergers and joint ventures.[72] Collaborative weapon development or production with foreign companies can spread development costs and risks, while pooling technical knowledge among allies. Collaboration with foreign firms has the drawback of transferring American defense technology to companies overseas and reducing opportunities for domestic production. The benefits and costs of cooperative efforts will need to be weighed in terms of their long-term effect on the production base and the national economy.

When systems either are not available or are more expensive on the domestic market, the United States can also place greater reliance on foreign sources of military supplies and components. This choice between materiel autonomy and increased interdependence raises many questions. In the extreme, the DoD could compete procurement contracts worldwide and take the best bid whatever its origin. Currently, foreign sourcing is restricted primarily to the lower tiers of the production base, although important subcomponents (e.g., flat-panel displays) and systems (e.g., the AV-8A Harrier jump jet, the M119 105mm howitzer, and the Berretta 9mm sidearm) have been purchased or produced under license. The full extent of foreign content is not well understood by the DoD because of the difficulty of tracking all the parts in a system.[73] Moreover, many large corporations generally regarded as American (e.g., IBM) are in fact international in scope and perspective.

The risks of foreign sourcing will have to be weighed against the cost of sourcing components and systems domestically.[74] One risk is a political cutoff of items that affect U.S. capabilities in a crisis or war.[75] Cutoffs could also result from a military blockade.[76] Stockpiling items that will not quickly become obsolete and multiple sourcing of foreign components can decrease vulnerability to a cutoff. Another risk is the potential for U.S. technological dependence on other countries. This dependence would not only affect current systems, but the capability to produce future systems as well. Creating protected industries to preserve an uneconomical capability against a product cutoff will cost the government more, reduce incentives to innovate, and constrain access to foreign technological advances. Since purchases overseas will deprive the U.S. industrial base of the dollars transferred abroad, Congress might consider loosely tying foreign military purchases to foreign military or commercial sales.[77]

[71] Unless bans are universally enforced, they will not prevent countries from obtaining weapons and may serve primarily to cut signatory nations out of the export sales market. If all the industrialized democracies participate in a ban, however, the level of weapon sophistication sold into unfriendly hands may decline.

[72] One journalist reported a trend in Europe towards a protected defense market. Trade barriers, offset agreements, and tariffs have been combined to create a Buy European atmosphere. Great Britain and the Netherlands are attempting to deregulate a community that in much of Europe is government-controlled or -owned. (Patrick Oster, "Europeans Shelving Rivalries Over Big Weapons Contracts: Possible Trend Concerns U.S. Defense Firms," *The Washington Post*, Sept. 11, 1991, pp. C1 and C3.)

[73] U.S. Congress, General Accounting Office, "Industrial Base: Adequacy of Information on the U.S. Defense Industrial Base," GAO/NSIAD-90-48 (Gaithersburg, MD: U.S. General Accounting Office, November 1989).

[74] Testifying before the House Armed Services Committee's Panel on the Structure of U.S. Defense Industrial Base, Nov. 1, 1991, E. Gene Keiffer proposed that instead of investing money to reproduce domestically what can be bought more inexpensively abroad, it would be wiser to invest in leapfrogging the foreign competition and produce a next generation of the item.

[75] For example, normal sandbags intended for use in the Persian Gulf War were too porous for the fine sands of the Saudi desert, allowing their contents to filter out overnight. The only bags on the international market made for this type sand were being distributed by a Dutch firm whose main supplier was Iraq. The General Services Administration instead turned to U.S. manufacturers, which in the end produced 71 million bags. ("The Finer Points of Sandbagging," *Parade Magazine*, Jan. 12, 1992, p. 14.)

[76] Purely domestic sources are not immune to production cutoff. A variety of factors (e.g., accidents, severe weather, strikes, or moral outrage) might result in a shutdown in production. One example is the loss of tritium production as a result of environmental and safety concerns. For a discussion of the risks of both domestic and international dependencies, see "An 'Adequate Insurance' Approach to Critical Dependencies of the Department of Defense," op. cit., footnote 8.

[77] At present, the United States sells far more military equipment overseas than it buys.

Global interdependence may also result in an increasing number of foreign acquisitions of U.S. defense companies. This situation may not be critical if the acquired companies continue to work for the DoD and obey export laws. During the Persian Gulf War, for example, Conventional Munition Systems of Tampa, a wholly owned subsidiary of the German firm Messerschmitt-Bölkow-Blohm, rushed U.S. Army orders for Patriot missile warheads and parts, in addition to manufacturing Maverick missile warheads.[78] If a foreign-owned firm opted not to assist the DoD in a crisis, it could be compelled legally to live up to existing contracts, or, in extreme cases, be nationalized. Moreover, such firms' workforces and infrastructures remain resident in the United States, although patents might be held abroad. If such a company decided to leave the defense business, it would be no different than any number of American-owned firms now doing so. The Committee on Foreign Investment in the United States reviews foreign acquisitions and warns the President of potential threats to national security.[79] The DoD can also restrict foreign investment in firms that do classified defense work.

Promote Manufacturing Innovation

The introduction of manufacturing innovations is another method to stretch limited defense procurement dollars. Capital investment, design, and production can be altered to reduce the life-cycle costs of a product.

If procurement funding were to be made more predictable in the future (e.g., through multiyear allocations), contractors' capital investments could be optimized for long-term production efficiency, particularly when initial production has not yet begun. Facilities, personnel, and manufacturing equipment could be sized appropriately for the job without the higher costs associated with overcapacity or the delays caused by undercapacity. And organizational structures could be adapted to fit new production realities. Many ideas have been proposed to modernize American manufacturing—largely as a response to foreign competition—that have relevance for defense manufacturers. These range from well-understood techniques and technologies that can be implemented immediately to futuristic visions that give manufacturers a sense of direction, but cannot soon be implemented.[80]

In the near term, manufacturers can increase their current reliance on computer technologies to manage resource allocations more efficiently (e.g., just-in-time supply or staffing) and communications with suppliers and customers (e.g., computer-aided acquisition and logistics).[81] Computer-controlled machine tools that can flexibly switch from manufacturing one item to another with a change of programing are already a common component in many factories and may become more prevalent as defense procurement moves away from high-rate production toward low-rate production.[82] Typically, these machines are limited to a few related tasks and do not manufacture a complete system. Organizational innovations, such as quality programs (e.g., Total Quality Management or Zero Defects Management) and working in group cells, can also be adopted.

Martin Marietta has adopted several of these innovative technologies and techniques in its Orlando, Florida, LANTIRN navigation pod production facility, where it has established what it terms a "paperless factory." The factory uses a centralized computer system to keep track of all elements of the production process from inventories to product testing. The computer system even displays step-by-step manufacturing process information through

[78] Stuart Auerbach, "U.S. Firms Angered by Kuwaiti Contract Award," *The Washington Post*, Oct. 29, 1991, p. D1. This article focuses on a controversy over awarding CMS a contract set aside for American businesses to carry out demolition work in Kuwait.

[79] *Report to Congress on the Defense Industrial Base*, op. cit., footnote 15, ch. 4, p. 5.

[80] The Iacocca Institute at Lehigh University is championing one such vision, which it calls "Agile Manufacturing." This manufacturing strategy for the next century, sponsored in part by the DoD Manufacturing Technologies (MANTECH) program, describes a wholesale renovation of traditional American manufacturing that emphasizes brainpower and flexibility. The goal is the creation of a world-class business organization and infrastructure that can rapidly design and produce small lots of high-quality and long-lasting custom products more economically than mass produced goods. See Roger N. Nagel and Rick Dove, *21st Century Manufacturing Enterprise Strategy*, vols. 1 and 2 (Bethlehem, PA: Iacocca Institute, Lehigh University, 1991).

[81] According to DoD documents, cost savings of up to 20 percent of a program's budget are possible through Computer-aided Acquisition and Logistics Support (CALS) systems alone. (Neil Munro, "Pentagon Urges Industry to Streamline With CALS," *Defense News*, vol. 6, No. 36, Sept. 9, 1991, p. 39.)

[82] In 1988, the U.S. Bureau of the Census reported that numerically controlled machines were used in 32-56 percent of the heavy industry businesses sampled. Elaborate information systems, such as CALS, were employed in significantly fewer companies. See *Statistical Abstract of the United States: 1991*, op. cit., footnote 69, p. 760.

video and animation at each employee workstation. Construction, however, remains primarily a hands-on process, with little reliance on robotics. According to Martin Marietta, the $40 million dollar investment reaped $100 million in cost savings in the first 4 years.[83]

Further in the future, already extant computer-aided design (CAD) capabilities will become increasingly integrated with computer-aided manufacturing (CAM). At first, this will mean making data packages from CAD systems readily convertible to CAM systems;[84] eventually, this process will be automated and directly linked. Then engineers will be able to draft their designs on the same computer system that will later direct man and machine through the manufacturing process. Late design changes and error corrections made by engineers will be transferred immediately throughout the factory, ensuring configuration and inventory control.

Initially, CAD/CAM systems will be used to make discrete components of a system, but not the system itself. A plant of the future, where artificially intelligent computers create the actual design of a product (as opposed to simply graphically representing a human design) and then task robots to manufacture it with limited human intervention, is well beyond current capabilities in most industrial sectors, but it is the target many innovators are working toward.[85]

Another means of improving defense production efficiency is *concurrent engineering*.[86] In concurrent engineering, the traditional sequential process of design, development, production, and maintenance is abandoned in favor of a more unified approach. Experts in manufacturing and maintenance are brought early into the design process to lend their expertise. This multidisciplinary team eases the normally rough transition from development prototype to production by emphasizing producibility at every step. (See figure 4-6.)

Figure 4-7 compares program spending over time for procurement contracts based on concurrent and traditional engineering. The curve corresponding to concurrent engineering rises earlier, reflecting the cost of including manufacturing engineers and maintenance personnel in the design process. In order for this "front-loaded" curve to pay off, the total cost of the program spread over the full production run (and over the entire product life cycle) must be less for concurrent engineering than would be paid traditionally. This reduction in overall cost comes from a smooth transition from development to production, which avoids many of the mistakes, waste, and delays common in traditional production runs, making manufacturing and maintenance easier. Development time can also be shortened through concurrent engineering, although in an era of tight budgets and reduced threats short cycle time may not be a high priority. Special attention needs to be applied to how concurrent engineering will fit into an acquisition strategy that emphasizes prototyping over production. (See ch. 3.)

Companies with a large proportion of commercial work are more likely to innovate in the manner described above. Defense contractors now dependent on the DoD, however, may need special incentives to innovate their manufacturing.

In all cases, **the benefits of improved manufacturing technologies and processes must be weighed against the costs of continuing defense production in the current fashion.** For many industries, particularly those that produce specialized products in small lots, automation may not be

[83] OTA site visit on Sept. 11, 1991; and Steven Pearlstein, "Contractors' New Watchword: Efficiency," *The Washington Post*, Dec. 11, 1991, p. A1 and A18.

[84] A 1985 DoD report stated that "a common database between design and manufacturing functions has inherent technical problems but has the highest potential payoff in product quality and productivity." Department of Defense, Assistant Secretary of Defense for Acquisition and Logistics, *Transition from Development to Production . . . Solving the Risk Equation*, DoD 4245.7-M (Washington, DC: Department of Defense, September 1985), ch. 5, p. 24.

[85] U.S. Congress, Office of Technology Assessment, *Computerized Manufacturing Automation: Employment, Education, and the Workplace*, OTA-CIT-235 (Washington, DC: U.S. Government Printing Office, April 1984), p. 6.

[86] In a study of concurrent engineering for the Office of the Assistant Secretary of Defense for Production and Logistics, analysts at the Institute for Defense Analyses came up with this definition of the term:

> Concurrent engineering is a systematic approach to the integrated, concurrent design of products and their related processes, including manufacture and support. This approach is intended to cause the developers, from the outset, to consider all elements of the product life cycle from conception through disposal, including quality, cost, schedule, and user requirements.

See Robert I. Winner et al., "The Role of Concurrent Engineering in Weapons System Acquisition," IDA Report R-388 (Alexandria, VA: Institute for Defense Analyses, December 1988), p. 2.

Photo credit: Martin Marietta

(Above) Engineers in a centralized control room oversee production in Martin Marietta's "paperless factory." (Below) A technician checks his work against a diagram on his computer terminal.

a cost-effective alternative to hand crafting and assembly. Moreover, it makes little sense to spend resources reducing direct labor in the construction of an item where labor costs are negligible compared to component costs (e.g., satellite assembly) or where industry is currently overcapitalized, unless such investment produces other gains (e.g., increased reliability).

Currently, defense firms are constrained in the adoption of manufacturing innovations by many factors, including the disincentive of annual contract renegotiations that eliminate profits achieved through increases in productivity. Until now, almost all the manufacturing innovation that has occurred has been evolutionary, stemming largely from contractor initiative, contract requirements (e.g., new composite material fabrication techniques on the B-2), independent research and development (IR&D), and through manufacturing technology programs sponsored by the DoD, such as the Manufacturing Technology Program (MANTECH) and the Industrial Modernization Incentives Program (IMIP). MANTECH programs fund manufacturing process, material, and equipment R&D. IMIP programs incorporate incentive clauses into contracts to motivate contractors to adopt proven manufacturing innovations that the contractors would not be able or willing to sponsor themselves. Increases in productivity, quality, and reliability are designed to benefit both the company and the government. MANTECH and IMIP funds have largely gone to prime contractors and not to subcontractors and suppliers.[87]

As the United States moves into an era of reduced defense procurement, many of the traditional sources of funding for manufacturing innovations are beginning to dry up. Many surviving defense firms will not be compelled by commercial market pressures to innovate and will have fewer procurement dollars to invest than in the past. Natural monopolies will have less incentive to update their manufacturing technology than companies forced to stay competitive. In these circumstances, Congress could fund MANTECH and IMIP programs as one approach to bringing innovation to these segments of the future base.[88]

Although manufacturing technology programs have been in existence since the 1950s, they have become much more prominent in recent years,

[87] John A. Alic et al., *Beyond Spinoff: Military and Commercial Technologies in a Changing World* (Boston, MA: Harvard Business School Press, 1992), p. 344.

[88] U.S. Department of Defense, Defense Systems Management College, *Defense Manufacturing Management: Guide for Program Managers*, 3d ed. (Washington, DC: U.S. Government Printing Office, April 1989), ch. 4, p. 5 and ch. 8, p. 5.

Figure 4-6—Decisions Concerning Producibility

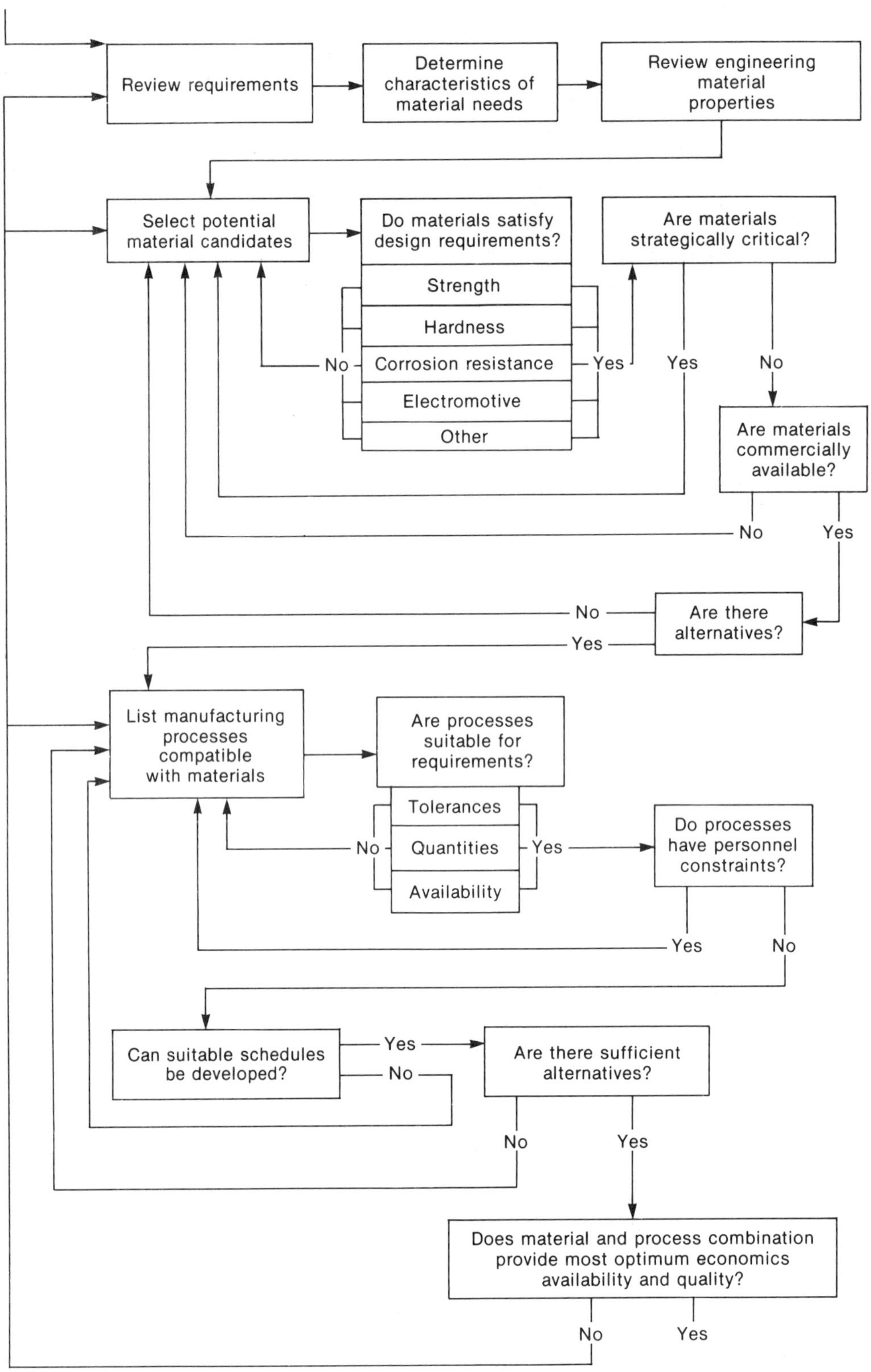

SOURCE: Defense Systems Management College, 1989.

Figure 4-7—Theoretical Comparison of Concurrent v. Traditional Development and Production Programs

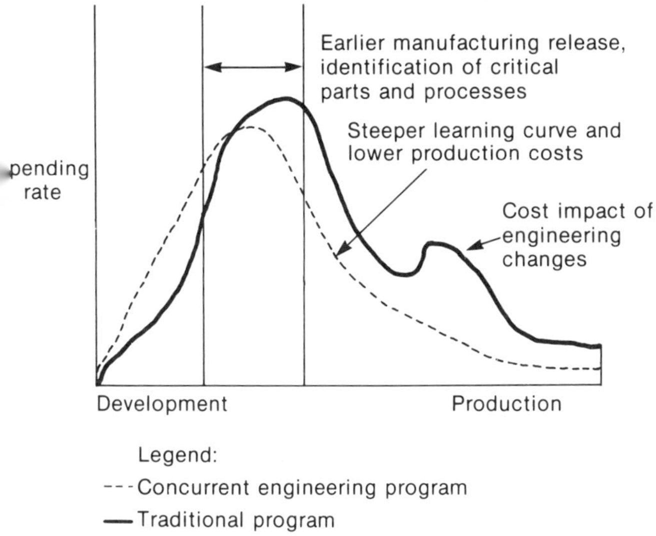

SOURCE: Office of Technology Assessment, 1992.

largely through congressional intervention.[89] Each of the Services runs a separate MANTECH program, in addition to a program run by the Defense Logistics Agency. These programs, while focusing on different segments of industry, strive to bring government, industry, and academia together (often in regional technology centers) to produce generic manufacturing technology innovations that can be transferred to the defense production base. MANTECH ventures have been successful in processing gallium-arsenide wafers for advanced microelectronics, in nondestructive imaging of products, and in robotic ship welding.[90] In addition, since 1985, the Navy's Best Manufacturing Practices program has sent survey teams to manufacturers to discover what they are doing right and transfer this knowledge to the rest of the Navy's production base.[91] The DoD has also used the Asset Capitalization Program, authorized by Congress in fiscal year 1983, to fund the modernization and acquisition of equipment for such operations as depots and shipyards. (See ch. 5.)[92]

A final method for increasing production base efficiency through manufacturing innovation would be to construct systems to incorporate modular subsystems so that when a subsystem is broken or needs to be upgraded, it can be readily replaced with a new, self-contained unit. The removed unit would then either be sent back to a depot or to the original equipment manufacturer (OEM) as part of a strategy to maintain the manufacturing capability of the OEM. Modular subsystems could be made common across several platforms (e.g., a plane, helicopter, and tank could all use the same radio) to generate economies of scale in production, and they would allow generic weapon platforms to be specially outfitted for different missions.

The primary drawback of switching to modular systems is that they may require built-in slots or boxes to hold them that would increase the overall cost and weight of the system, resulting perhaps in lower performance. For example, it has been suggested that many naval ships could be built according to one basic hull design that would accept a variety of weapon and equipment modules according to its mission (e.g., antisubmarine warfare, air defense, cargo, or amphibious assault).[93] While containers for these modules might add as much as 5 percent to the cost of a single ship, the economies of producing identical ship hulls might result in lower total cost.[94] Other modular systems might include the next-generation tank or multirole fighter. The potential added cost and reduced optimization of modular systems will have to be weighed against

[89] Congress has repeatedly authorized more funds for MANTECH than the DoD has requested. This is due in part to a difference in perspective. The DoD, along with the Services, sees MANTECH as a tool for increasing productivity for specific weapon systems. Congress, on the other hand, sees MANTECH as a stimulus to the production base as a whole, including the nondefense community.

[90] Torelli et al., testimony before the U.S. Congress, House Armed Services Panel on Future Uses of Manufacturing and Technology Resources, Oct. 24, 1991; and *Computerized Manufacturing Automation,* op. cit., footnote 85, pp. 314-316.

[91] Torelli, ibid; and Department of the Navy, *Best Practices: How to Avoid Surprises in the World's Most Complicated Technical Process, The Transition from Development to Production,* NAVSO P-6071 (Washington, DC: U.S. Government Printing Office, March 1986).

[92] U.S. Congress, General Accounting Office, "Industrial Funds: The Department of Defense's Management of ACP Funds," GAO/NSIAD-90-202FS (Gaithersburg, MD: U.S. General Accounting Office, June 1990); and U.S. Congress, General Accounting Office, "Plant Modernization: DoD's Management of the Asset Capitalization Program Needs Improvement," GAO/NSIAD-89-147 (Gaithersburg, MD: U.S. General Accounting Office, August 1989).

[93] The German shipbuilder Blohm and Voss constructs two classes of modular ships: the Meko and the Deutschland (Type 123) frigates.

[94] About 70 percent of the cost of a modern warship is for systems, such as weapons and electronics, other than the hull. See Gary Hart and William S. Lind, *America Can Win: The Case for Military Reform* (Bethesda, MD: Adler and Adler, 1986), pp. 98-107.

Figure 4-8—Tradeoffs Between Production Efficiency and Surge and Mobilization Preparation

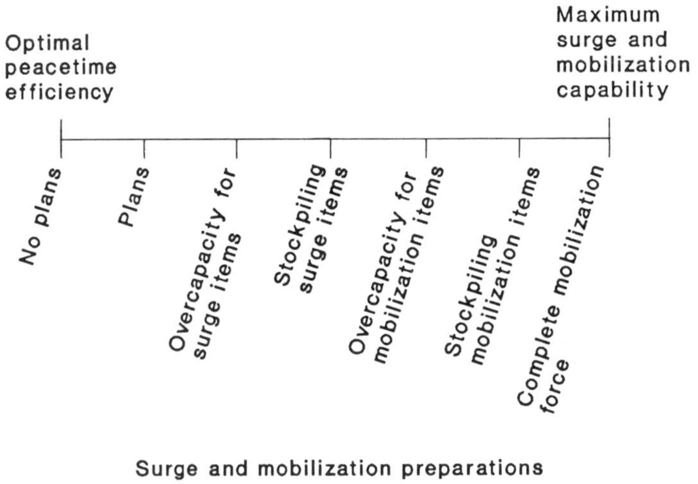

SOURCE: Office of Technology Assessment, 1992.

lower production and maintenance costs and higher operational readiness. (Ch. 5 discusses maintenance issues in greater detail.)

Increase Commonality

Increased commonality in Service procurement and among defense products could make the base more efficient. One notion is the creation of a more common procurement process among the armed services. The aim would be to reduce redundant procurement programs and to make larger purchases to achieve economies of scale. (Ch. 6 discusses options for rationalizing the procurement structure in more detail.)

The Services could also try harder than they currently do to build new systems with a greater emphasis on common components and standardized parts. The Army is attempting greater standardization of parts and systems through such programs as the Armored Systems Modernization program which is designing the next generation of armored vehicles.[95] This program is being redefined, stretched out, and scaled down in the face of budget reductions. Current plans call for the development and production of the Advanced Field Artillery System and the Future Armored Resupply Vehicle-Ammunition on a heavy-level protection chassis and the Line-of-Sight Antitank system on a medium-level protection chassis. Three other vehicles intended for the heavy-level protection chassis have been deferred indefinitely: the Block III tank, the Combat Mobility Vehicle, and the Future Infantry Fighting Vehicle. In addition to the common chassis, these armored vehicles will have significant commonality in components, including armor, engines, tracks and suspension, electronics, instruments, hardware, and software. Although the parts for the vehicles will not be completely interchangeable,[96] they will be as compatible as their differing missions will allow.[97] More universal parts would help those subcontractors that might go out of business if their product supported only one weapon system.

In addition to the above steps, the DoD should examine how much standardization of parts, munitions, and systems there should be internationally, especially with our NATO allies. One of the DoD's

[95] A General Motors Corp. study of an earlier armored vehicle modernization plan developed several manufacturing options that emphasized commonality in construction, vendors, and parts. See General Motors Corp., Military Vehicle Operations, "Manufacturing Plan Appendix," *Armored Family of Vehicles (AFV) Phase I Study*, 1986.

[96] John G. Roos "$59-Billion Armor Mod Plan Has Only One Tread on the Ground," *Armed Forces Journal International*, October 1991, p. 60.

[97] An example of the increased efficiencies of commonality is found in the Mazda Miata. This petite and sassy sports car—the automotive sensation of 1989—appeared completely different from other Mazdas, but was built with 80 percent standard parts. This strategy allowed Mazda to bring a new product to market quickly and make a profit despite low volume sales. (Peter F. Drucker, "The Big Three Miss Japan's Crucial Lesson," *The Wall Street Journal*, June 18, 1991, p. A18.)

long-standing objectives has been to promote the adoption of standardized or interoperable equipment among allies and friendly nations. This issue should be examined from the perspective of cost savings, base support, foreign dependence, and the changed international environment.

RESPONSIVE PRODUCTION

A future crisis might require the production base to react through either a responsive surge of the defense base or a longer term mobilization of the broader national industrial base.[98] The next two sections discuss the balance between peacetime production efficiency and crisis requirements. Figure 4-8 illustrates some tradeoffs between efficient peacetime production and surge and mobilization preparations.

Policymakers can make the future base responsive to crises short of a national emergency in three ways: by surging production as required by commanders, by stockpiling products in advance, or by relying on allies. Each of these options has advantages and disadvantages.

If production is sufficiently responsive, then the government does not need to pay for surge items unless there is a crisis, nor does it have to pay for storage. Relying on surge carries the risk that items cannot be produced quickly enough to meet the field commanders' needs. Moreover, surge facilities may entail higher overhead costs by maintaining more production capacity than is needed for peacetime requirements.[99]

If items are stockpiled, they are available on demand if ever needed, but at a high up-front cost, and they may not be replaceable if production facilities close and requirements surpass stocks. Some items, like electronic components, become obsolete so quickly they are not conducive to long-term storage.[100] A rolling production inventory—an early buy of components to be used in final production items—might reduce some of these costs.

Table 4-4—Examples of Surge and Mobilization Items

Surge items	Mobilization items
Ammunition	Surge items (from list at left)
Food	New weapon systems
Fuel	Mothballed weapons
Uniforms	Commercial transport
Spare parts	Commercial engineering vehicles
Medical supplies	National Defense Stockpile materials
Merchant marine	
Prepositioned equipment	

SOURCE: Office of Technology Assessment, 1992.

Foreign acquisitions have the possible advantage of lower cost, but run the risk of political cutoff and are less likely to be able to meet the quantity requirements of the U.S. military. Foreign items may also suffer from excessive transportation lag times.

Planners should use contingency plans for future crises to designate which items should be procured in advance, which should be surged, and which should be obtained from allies. The resulting system will need to be properly funded and exercised periodically to ensure it will work when needed.

It is unlikely that production of major weapon systems will have to be surged for a conflict that falls short of a national emergency. Moreover, if the United States pursues low-rate production, the surge of such systems will be virtually impossible. It is more likely that field commanders will need increased production of consumable or personal items, such as munitions, spare parts, fuel, food, and clothing. (See table 4-4.)

The DoD interim report to Congress on the conduct of the Persian Gulf War provides some

[98] U.S. Congress, Office of Technology Assessment, *Adjusting to a New Security Environment: The Defense Technology and Industrial Base Challenge*, OTA-BP-ISC-79 (Washington, DC: U.S. Government Printing Office, February 1991), p. 4. This report maintains the definition used in *Adjusting to a New Security Environment*. *Surge* is the term used within the DOD to refer to the expansion of military production in peacetime without the declaration of a national emergency. *Mobilization* refers to the rapid expansion of military production to meet material needs in a war and involves the declaration of a national emergency. Several types of mobilization are considered. *Full mobilization* refers to mobilization to fill the existing or "program force" structure. *Total mobilization* describes a mobilization effort that expands beyond the existing force structure. Mobilization is often referred to as "reconstitution" by the current Administration.

[99] Plants designed for efficient peacetime production can expand their work hours, at least temporarily, if they are not already operating at maximum capacity. Longer term reliance on extended or additional work shifts will require the hiring of skilled or trainable personnel.

[100] The market lifespan of an electronic component has decreased from 10-12 years to 4-5 years, while weapon system longevity is 20 years or more. U.S. Congress, General Accounting Office, "Defense Inventory: DoD Could Better Manage Parts with Limited Manufacturing Sources," GAO/NSIAD-90-126 (Washington, DC: U.S. Government Printing Office, August 1990), p. 8.

useful examples.[101] Generally, the Services had the major equipment and supplies they needed before the crisis, but shortages of some items soon emerged. T-rations, designed to feed 8-10 people, had not been included in the war reserves and, for a time, industry could not meet the increased demand. Troops were temporarily forced to eat the less palatable Meals, Ready-to-Eat (MREs), which had been stockpiled. Many troops also were initially stationed in Saudi Arabia dressed in uniforms camouflaged for European woodlands, while the clothing industry manufactured clothing and boots patterned for the Kuwaiti desert. While both of these shortfalls caused problems, they did not significantly impede operational preparations. Shortages in the U.S. inventory of heavy equipment transporters and offroad vehicles were compensated for by leasing, buying, or requesting donation of trucks from U.S. trucking companies, and from Saudi Arabia, Germany, Egypt, Italy, and Czechoslovakia.

According to the report:

> Literally thousands of items were accelerated to meet the increased requirements of U.S. Central Command (CENTCOM). From weapons systems to individual items of supply, a tremendous demand was placed on the nation's industrial base. Items such as chemical protective clothing were surged from 33,000 per month to 70,000 per month, desert combat boots went from zero to 124,000 per month, and desert camouflage uniforms went from zero to 376,000 per month over a six month period. In some cases, the increase in the production rate was the direct result of an individual contractor's performance, in other cases, additional contracts were required. Preliminary investigation indicates that despite some shortcomings, the industrial base was reasonably responsive to the needs of the force. These and similar instances reinforce the continuing requirement to balance our war reserve programs and depot production capabilities with a realistic assessment of industrial base capability.[102]

Extensive preparation time, control over the timing of operations, a short war, relatively light combat damage, support from Saudi Arabia and the other coalition partners, and a lack of a major threat elsewhere made the Persian Gulf surge effort easier than it might have been otherwise.

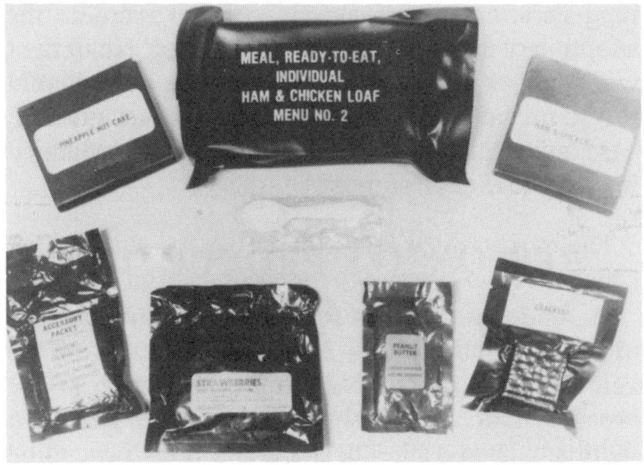

Photo credit: The DoD

Meals, Ready-to-Eat (MREs) are nutritious and energetic foods packaged to survive the rigors of a combat environment.

Flexible manufacturing systems, besides being useful for an efficient peacetime base, can affect future responsiveness. On the positive side, they will make it easier for companies engaged in peacetime defense work to shift from a lower to a higher priority product mix (e.g., from dress to camouflage uniforms) to meet specific surge demands.[103] Companies that produce both commercial and defense products would be able to temporarily halt commercial work and expand defense production. However, a production line set up for flexible manufacturing—where excess capacity has been cut to the bone—may make any expansion of production more difficult if the majority of a company's products are required for surge.

PREPARATION FOR MOBILIZATION

If a future crisis is severe enough to warrant a declaration of a national emergency, the surge capability described above may not be adequate to meet the challenge. Full or total mobilization (currently dubbed "reconstitution") of the broader national industrial base, in addition to the defense production base, may be necessary. With the disappearance of the Warsaw Pact threat to NATO in Europe, a war on this scale seems unlikely for the

[101] *Conduct of the Persian Gulf Conflict*, op. cit., footnote 53, ch. 7, pp. 1-7. This report carries the caveat that its information is preliminary and subject to change.

[102] Ibid., ch. 7, p. 2.

[103] The time it takes to shift from one product to another will depend on the degree of tool flexibility and product similarity.

foreseeable future, so it appears prudent to give more priority to peacetime and surge planning. Nevertheless, **realistic planning for mobilization against a major threat remains essential to the security of the Nation and its allies, especially considering the long lead times involved.**

The most critical factor in mobilization planning is the amount of warning time the industrial base will have. This warning time depends on the speed of an adversary's mobilization, the timeliness and reliability of intelligence, the length of time ready forces can hold their own before being reinforced with mobilized reserves and supplies, and the time required for a political decision to mobilize. The national industrial base would have about 2 years warning of a major war in Europe, according to current projections.[104]

Overestimating warning time in the planning phase can lead to serious shortages in the early stages of a war. Underestimating warning time can lead to an overinvestment in stockpiled supplies and too little investment in manufacturing resources, leading to a full inventory at the beginning of the conflict, but a declining capability as it proceeds. Improved planning tools based on detailed production data and models can help prepare for large-scale mobilization, but only if the subjective inputs of crisis scenarios are accurate.[105]

Once planners have made their best estimate of mobilization warning times, they can decide the best way to meet mobilization requirements. Equipment that cannot be produced within the warning time must be stockpiled in the national War Reserve or obtained from U.S. allies. Other items might also be stockpiled, but as mentioned above, stockpiling involves a large up-front investment in equipment and supplies that may never be used. Moreover, the military may have difficulty replacing stocks if demand has been underestimated or after the crisis is over. Products that can be manufactured within the warning period but exceed the surge capacity of the defense production base will need to be procured through an expansion of the defense sector or from the commercial sector.

Photo credit: BMY-Combat Systems

The flexibility of a computer-controlled machining center permits BMY to move in minutes from working on aluminum armored vehicle hulls to steel hulls, such as the M88 Medium Recovery Vehicle pictured here.

The dedicated defense production base provides a core around which the civilian base can be mobilized. The existence of a solid core of personnel, equipment, and facilities will depend on what measures policymakers take now during peacetime procurement cutbacks (e.g., none, creation of an arsenal, low-rate production, or civil-military integration). The implementation of rapid acquisition rules and simplified procurement procedures after the declaration of a national emergency will allow existing facilities to be expanded and new ones to be built. Facilities that would require long lead-times to build and outfit may have to be mothballed in peacetime (e.g., shipyards).[106] Companies that have been participants in mobilization planning and those

---

[104] *Redesigning Defense*, op. cit., footnote 1, p. 24.

[105] For example, the Army is funding a prototype Production Expansion/Acceleration Capability Enhancement (PEACE) computer model developed by Salvador Culosi at the Logistics Management Institute, which optimizes funding for a particular product and its subcomponents to meet peacetime and mobilization requirements based on such inputs as likely crisis scenarios, plant capacity, industrial planning measures, process flow times, and product and critical subcomponent lead times.

[106] Mothballing facilities may not be a good option for maintaining a production capability if similar work is not being performed elsewhere. For example, a company that currently produces ammunition or armored vehicles might be able, with significant difficulty, to bring on-line another mothballed munitions plant or tank facility, but it would face greater, if not insurmountable (in the time provided), difficulties in restarting production if it had not manufactured the product for several years. The Canadian Navy recently encountered this dilemma when they tried to construct a frigate without having built one in a decade. The result was a substantial expansion of costs and schedule and the need to rely heavily on foreign expertise. If it is necessary to mothball an entire capability for financial reasons, then every effort should be made to document production procedures and worker experience (perhaps by creating computer expert systems) before they are lost.

Table 4-5—Comparison of Production Issues

Desirable characteristics of the future production base	Policy options: Streamline and consolidate	Enhance government and industry cooperation	Foreign cooperation	Reduce production rates	Promote innovation
Efficient	Reduces overcapacity, preserves vital production capabilities	Lowers product cost, allows defense base to draw on national base, makes defense work more appealing, but risks abuse	Expands sales market, source of technology and components, but undermines domestic base	Short-term efficiency loss, but maintains production capabilities efficiently after streamlining	Raises efficiency, but may demand capital investment
Responsive	Maintains production capabilities, but decreases capacity	Cooperative atmosphere eases planning for future	Expands domestic base, but risks cutoff	Maintains production capabilities, but reduces capacity	Flexibility allows surge of select items, may reduce capacity
Mobilizable	Maintains production capabilities, but decreases capacity	Cooperative atmosphere eases planning for future	Essential for large-scale conflict, but risks cutoff	Maintains production capabilities, but reduces capacity	Future flexibility may allow easier transfer of production to national base

SOURCE: Office of Technology Assessment, 1992.

that are flexibly organized (e.g., with working group cells) will be better prepared to make the transition to mobilization.

The health of the national economy is vital to a successful mobilization. In a national emergency, the DTIB will need to draw extensively on the skills, facilities, and management of nondefense manufacturers. The commercial sector can be relied on for a large number of off-the-shelf items or items that are easy to adapt to military standards. This will be particularly true if steps are taken now to integrate the DTIB with the broader civil production base. Mobilization of militarily unique systems, such as armored vehicles and airplanes, will require preplanning by mobilization planners with the cooperation of defense manufacturers and the retention of critical manufacturing skills and equipment.

In the future, the spread and standardization of flexible computerized manufacturing tools throughout industry might make it easier to switch in a national emergency from commercial to defense production.[107] This might be especially true in factories that produce both military and civilian items on the same equipment. During mobilization, designated companies would cease commercial production and use that freed capacity to manufacture military items. The DoD can foster such a future by lowering the barriers to civil-military integration (as described earlier) and through manufacturing technology programs that emphasize equipment and data-format standardization. For example, a government-funded model factory or laboratory might be established to design machine-tool data packages and software for manufacturing weapon components that could be transferred easily to flexible commercial plants in the event of a mobilization. Currently, manufacturers in both the commercial and defense industries lack this degree of flexibility, but the necessary technologies are emerging and might be fostered with the right incentives.

The industrial bases of U.S. allies will also be an integral part of any future mobilization effort. The magnitude of such a crisis would demand some division of labor among allies, regardless of the risk of material cutoff. Promoting mutual defense cooperation with allies and friendly nations, protecting

[107] Flexible manufacturing will also make it easier for pure defense producers to switch from lower to higher priority end items. For example, the Armored Systems Modernization program mentioned earlier, which planned six armored vehicles based on two common chassis, is being designed for more flexible manufacturing. The two chassis will be manufactured on the same assembly line using numerically controlled machine tools. Mission packages in more or less modular form will permit relatively rapid shifts from producing one type of vehicle to some other vehicle that is in higher demand. Unlike a shift from commercial to defense production, however, flexibility in strictly defense firms alone will not expand their total military output, only priority items (e.g., a tank rather than an artillery vehicle). See Roos, op. cit., footnote 96, p. 60.

Table 4-5—Comparison of Production Issues (continued)

Policy options (continued)					
Support manufacturing skills	Increase common system procurement	Increase component commonality	Maintain surge capability	Stockpile	Lay-away facilities
Fills critical gaps	Reduces product life-cycle costs, but products are less mission oriented	Lowers costs and concentrates manufacturing	Raises overhead costs	Allows short-term production, but product may never be needed	Cost without product
Provides labor pool for base expansion	Eases logistics	Eases logistics	Allows increased production on short warning	Products available when needed, reduces foreign dependency, may be irreplacable once used	Eases expansion of production, particularly for facilities that are difficult to rebuild
Provides labor pool for base expansion	Eases logistics	Eases logistics	Allows expanded defense base production while national base gears up	Products available when needed, reduces foreign dependency, may be irreplacable once used	Eases expansion of production, particularly for facilities that are difficult to rebuild

global sea lines of communication, and, perhaps, maintaining a forward presence, would help ensure the viability of such overseas collaboration.

CONGRESS AND THE PRODUCTION BASE

Deciding on the necessary steps to restructure the defense production base will challenge many past notions about how the base ought to be run. Congress, in cooperation with the Executive Branch, will need to reevaluate many controversial issues, such as the relative importance of competitive procurement, contractor accountability, and buying American. Efficient peacetime production will have to be balanced against potential surge and mobilization requirements.

The measures Congress adopts during the transition to the future production base will depend on how damaging it believes procurement reductions of 50 percent or more will be to the Nation's defense industry. If Congress believes that production base problems will be limited to select industries with the rest able to adapt successfully to the new enviroment, it will opt for small adjustments to existing laws and practices. If it views the problem as more severe and fundamental, it may opt for a general restructuring of the production base and defense procurement. In either case, policies will need to be sensitive to the complexities of the base, particularly the different industrial sectors and tiers. Table 4-5 summarizes the measures discussed in this chapter. Below, these measures are discussed in groupings of particular interest to Congress.

Funding Decisions

Congressional control over DoD procurement funds and the rate at which these funds are reduced will have the most direct impact on the production base. Thoughtful reductions can ensure that future military requirements will be met. Greater consistency in procurement projections, perhaps with multiyear contracts, will allow the production base to reorganize more efficiently and manufacture defense equipment effectively at lower rates. Congress might further the efficiency of the future base by providing additional funds for the study of the base (e.g., composition and effect of laws and regulations), the adoption of manufacturing innovations, and the maintenance of critical manufacturing skills. Funding will be necessary for long-range planning, stockpiling, and the maintenance of excess peacetime production capacity in select areas to meet potential surge and mobilization requirements.

Base Structure

Congress has a range of options for restructuring the future production base. At a minimum, Congress should insist that the DoD identify critical producers at the supplier, subcontractor, and prime contractor levels that are at risk due to procurement reductions, and use existing laws and regulation to save their core capabilities. Public or private arsenals could be established for those industrial sectors that can no longer maintain themselves through DoD contracts.

Next, Congress could support government intervention in the market, if necessary, to save and strengthen critical producers through a combination of sole-source production, prototyping, upgrade, spare part, and maintenance contracts. For the good of the production base, the government might pick "winners and losers" or substitute "best value" for the lowest bid as the basis for awarding contracts. Congress could act to lower legal and regulatory barriers to mergers, strategic partnerships, and the creation of monopolies (e.g., antitrust laws and CICA) that undermine the consolidation of the base around select quality producers.

At the extreme, Congress might remove the legislative barriers to a full integration of the civil and military production bases, thus drawing on the size, efficiency, and innovation of the larger national base. After a time, only the most unique military items would remain in a separate DTIB, perhaps in arsenals (e.g., nuclear submarines).

Business Environment

Short of complete integration, Congress can act to relieve industry of many stifling characteristics of current defense work. These characteristics include costly paperwork requirements from bidding to final accounting, pervasive government oversight, outdated and obsolete specifications on many aspects of production, and a potential loss of a firm's competitive edge through the transfer of proprietary data rights to the government. The present business environment makes the production base inefficient and uninviting to innovative companies interested in doing defense work. Next January, a congressionally mandated DoD advisory panel will present its findings on how to streamline current acquisition laws.[108] Congress can act on the findings of this panel and of this report to foster a less adversarial relationship with industry. It can also encourage ongoing DoD efforts to procure more commercial products.

Acquisition Strategy

Congress could also promote the simplification and consolidation of the production base by supporting the consolidation of acquisition programs and organizations. It might also support commonality and modularity in weapon systems and subsystems and the use of multi-Service procurement to provide a more economic workload for a smaller number of core manufacturers. The government might also consolidate procurement efforts.

International Change

Finally, Congress needs to consider the role that the international defense production community will play in the future domestic base. The internationalization of the domestic base is already a reality, particularly at the lower tiers. Congress can act to increase this interaction by promoting military sales, purchases, and cooperative ventures; or it can opt to sever some or all of these ties, relying more on American industry.

SUMMARY

This chapter has discussed alternative policies to ease the transition to a future production base that has the desirable characteristics of efficiency, responsiveness, and the ability to mobilize. A thoughtful, orderly restructuring of the defense production base, in the face of a reduced international threat and pressing domestic financial concerns, is one of the biggest challenges facing defense policymakers. If the Administration and Congress do not take measures in the next few years, market forces combined with reduced defense spending will perform this restructuring haphazardly, resulting in a smaller, weakened, and potentially crippled DTIB. Some firms would weather these changes and continue to manufacture defense products. Others would be forced into other business areas or close. More than likely, should the need ever arise to surge capacity or mobilize, the United States would find itself lacking in critical capabilities.

[108] See footnote 34.

Chapter 5

The Maintenance Base

Contents

Figures

Tables

Chapter 5
The Maintenance Base

INTRODUCTION

The maintenance base is the third principal element of the defense technology and industrial base (DTIB). It is the portion of the base that supports deployed military systems, ensures force readiness, and sustains forces during military operations. *Redesigning Defense* discussed why a robust defense maintenance base will be vital in the national security environment the Nation faces in the future.

Defense maintenance is currently divided into at least three levels. The first is *organizational level maintenance*, where members of the operational unit make functional checks and adjustments, and faulty parts are serviced or replaced. The second is *intermediate level maintenance*, where field personnel perform more extensive repairs. The third is *depot level maintenance*, where highly trained personnel rebuild, make complex repairs on, and overhaul equipment in specialized facilities.[1] This chapter concentrates on depot level maintenance and uses the term "maintenance" to refer to it.

The U.S. military spent approximately $13 billion on depot level maintenance in fiscal year 1991, supporting a huge fleet of aircraft, armored vehicles, and other weapon systems and support equipment. (See table 5-1.)[2] Depot maintenance is currently the responsibility of the individual Services.

The depot maintenance system consists of two components. The **organic (i.e., Service-owned and operated) component** is composed of Army Depots, Air Force Air Logistics Centers, Naval Aviation Depots, Naval Shipyards, and Marine Corps Logistics Bases. This **in-Service** maintenance component employs about 150,000 people. It is supported by the **private sector** through the work of thousands of firms, including both repair houses and original equipment manufacturers. These firms supply parts and provide direct maintenance support in their own facilities or in government-owned and contractor-operated (GOCO) facilities.

Maintenance differs from production in that equipment arriving at the depot or factory for repair or overhaul must first be inspected and faults diagnosed. Once major items, such as ships, are disassembled to begin an overhaul, unanticipated repair requirements may be found, resulting in additional costs. Nevertheless, some maintenance and production activities are similar and many of the same skills are involved in manufacturing new parts or repairing old ones.

Future maintenance requirements will differ from those of the past 40 years. For example, the United States is likely to retire many weapons in response to the waning threat from the former Soviet Union and to arms control agreements. Since the oldest weapons are likely to be retired first, not only will the number of systems in the forces decline, but deployed weapons will initially tend to be newer and hence will require less maintenance. While some facilities such as shipyards are likely to have increased activities during the transition to a smaller force (e.g., decommissioning work), overall maintenance requirements will drop substantially.

Current trends, however, indicate a major reduction in new weapons procurement in the future. Thus, once present forces are reduced, the Nation will probably retain weapons and equipment in inventory longer than in the recent past, preferring to upgrade deployed systems when possible instead of producing new ones.[3] (See table 5-2.) This aging equipment may require more maintenance to retain high readiness levels. Also, military systems are becoming more sophisticated; in particular, the embedded electronic components are becoming more important and more complex. These trends will change the types of facilities needed for repairs

---

[1] The Army and the Marine Corps have five levels of maintenance: user, organizational, divisional, intermediate, and depot.

[2] The FY 1992 U.S. global military force, for example, includes over 45,000 armored vehicles, 490 combat ships, 4,100 major fixed-wing aircraft, and 260,000 Army tactical wheeled vehicles.

[3] The Congressional Budget Office noted that the expected changes in age of equipment are mixed, depending on the type of weapon. Between 1991 and 1995 ships will be relatively newer as will Air Force tactical aircraft. The average age of Army equipment and the age of Navy aircraft will increase. Statement of Robert F. Hale, Assistant Director, National Security Division, Congressional Budget Office before the U.S. House of Representatives, Committee on Armed Services, Mar. 19, 1991. After 1995, all classes of fielded equipment are likely to be older. Upgrading and retrofitting existing equipment is more similar to manufacturing than is repair, but such activities often take place in the maintenance, rather than in the production base.

Table 5-1—Depot Maintenance (fiscal year 1991 millions of dollars)

	Army	Navy	Air Force	Marine Corps	Total
Aircraft	294	1,830	4,001	a	**6,125**
Ships & boats	7	3,936	NA	NA	**3,943**
Combat vehicles	628	NA	NA	34	**662**
Missiles	190	60	278	12	**540**
Communications & electronics	244	12	70	10	**336**
Ordnance, weapons, munitions	53	150	19	3	**225**
Automotive	142	32	11	185	**370**
Other	48	675	300	36	**1,059**
Totals	**1,606**	**6,695**	**4,679**	**280**	**13,260**
Contract to private industry	34%	19%	27%	2%	**24%**

a Maintenance performed by Navy.
KEY: NA = not applicable.
SOURCE: Office of the Secretary of Defense.

and the skills of the people that perform maintenance work.[4] Some future upgrades and retrofits will aim at increasing the reliability of deployed systems, thus potentially reducing future maintenance workloads. For example, Rockwell International's current upgrade of F-111 avionics aims at improving reliability and maintainability.[5]

Force reductions and increased equipment reliability have already caused reduced workloads and overcapacity in the Service's present depot maintenance system. Future defense maintenance base objectives include:

1. preserving appropriate maintenance capability while forces are being reduced;[6]
2. providing maintenance support in peace, crisis, or war to a force that is likely to consist of older platforms that have been upgraded; and
3. supporting fewer but more sophisticated systems over the longer term.

Integrating more maintenance activities into the production element of the DTIB has been suggested as a way to sustain the defense production base and manufacturing skills in a period when less new equipment is produced. If this objective is accepted, it will have a significant effect on the size of the in-Service component of the maintenance base.

This chapter describes the current defense maintenance base, defines what is needed to have a robust maintenance base in the future, discusses some of the alternative ways of achieving a robust base and their policy implications. The options available for the maintenance base are similar to those in the production base. These options include:

Table 5-2—Average Age in Years for Selected Military Equipment

Equipment	1990	1993	1995
Air Force tactical aircraft	10	8	10
Navy combat aircraft	12	13	15
Naval surface combatant ships	15	13	14
Attack submarines	14	14	14
Ballistic missile submarines	18	15	11
M1 tank	NA	6	8
Bradley fighting vehicles	NA	6	8
M-109 howitzer	NA	23	25
AH-64 attack helicopter	NA	6	8

NA = not available.
SOURCE: OTA, based on information from the Congressional Budget Office, the Department of Defense, and the Department of the Army.

1. consolidating and restructuring the base while retaining its current character,
2. increasing use of the private sector,
3. increasing competition among Service organizations (depots and air logistics centers) and between Service organizations and private firms,
4. exploiting new technology, and
5. providing maintenance upgrades to U.S. equipment abroad, as well as foreign manufactured equipment.

[4] Alfred H. Beyer and Connelly D. Stevenson, *Depot Maintenance in the 1990's* (Bethesda, MD: Logistics Management Institute), July 1986, p. 4.

[5] William B. Scott, "Manufacturers Embrace Upgrades to Survive in '90s," *Aviation Week & Space Technology*, July 22, 1991, pp. 4-5.

[6] Defense Depot Maintenance Council, *Corporate Business Plan FY 91-95*, December 1991. The Council, for example, states that the "depot maintenance community finds itself faced with the challenge of having to downsize while simultaneously increasing efficiency and productivity in order to sustain forces in the field" in operations such as Desert Storm.

Photo credit: Rockwell International

U.S. and Australian F-111 fighter-bombers are upgraded side by side in this industrial facility. Maintenance of foreign-owned equipment could help support the U.S. maintenance base.

These options are not mutually exclusive but might be used in combination as a part of an overall maintenance strategy.

THE CURRENT DEFENSE DEPOT MAINTENANCE SYSTEM

In the past, each military Service has maintained its own equipment with the exception of a few select items (e.g., some aircraft engines), for which a single Service has assumed overall maintenance responsibility. The Services have traditionally sought ownership and control of maintenance for their own systems to ensure that they have the technical competence and resources to respond to emergency requirements.[7] The Services have also been concerned that failure to develop in-Service maintenance capabilities might leave them hostage to escalating cost demands by sole-source private contractors, or without the necessary support if the private sector determines that maintenance work is no longer profitable and leaves the business. However, these in-Service maintenance capabilities are expensive. For example, the acquisition by the Services of standard test program sets, which allow the military to test and repair complex electronics, can add up to 20 percent to the total development cost of a single electronics package. This cost would not be incurred if maintenance remained the manufacturer's responsibility. The increased use of the private sector is discussed later in this chapter.

Before fiscal year 1983, Service depots competed for equipment funds from the same pool that was used to acquire ships, aircraft, and other weapons systems; in many cases they were unsuccessful in obtaining funds to modernize their facilities.[8] During the expansion of the 1980s, however, the Service depots underwent substantial modernization funded by the DoD Asset Capitalization Program. The Services spent hundreds of millions of dollars on new equipment and in some cases replicated capabilities that already existed in another Service or in private industry.[9] By the end of the decade, however, the waning Soviet threat produced almost universal agreement that the existing capacity in the depot maintenance base exceeded future needs.

The Defense Management Report (DMR) released by Secretary of Defense Richard Cheney at the beginning of the Bush Administration identified ways to improve the management of the DTIB, including maintenance. Deputy Secretary of Defense Donald Atwood subsequently directed the Service Secretaries to prepare plans to reduce the cost of depot maintenance operations between fiscal year 1991 and fiscal year 1995 by a total of $1.7 billion "through internal streamlining and reducing the size of their maintenance depot infrastructure."[10] Among the specific actions directed were: transfer of workloads, establishment of one naval aviation depot maintenance hub on the east and one on the west coast of the United States, single-siting maintenance, improvement of labor productivity, and consideration of the withdrawal of Air Force maintenance activity from one of that Service's five main Air Logistics Centers. Another $2.2 billion was to be saved through long-range efficiencies that

[7] Kelvin K. Kiebler, Larry S. Klapper, and Donald T. Frank, *Army Depot Maintenance: More Effective Use of Organic and Contractor Resources*, report no. AR803R1 (Bethesda, MD: Logistics Management Institute, June 1990), p. 1-1.

[8] The military Services primarily use annual appropriations to reimburse the depots for actual work performed. Organic depots do not receive direct appropriations for this purpose; instead, they are funded indirectly using working capital in the Defense Business Operations Fund and orders from their customers to finance the cost of goods and services.

[9] Beyer and Stevenson, op. cit., footnote 4, pp. 7-9. The Logistics Management Institute study reported, for example, that the Army developed capabilities in microelectronics, automatic test equipment, and software that already existed elsewhere in the DoD.

[10] Deputy Secretary of Defense Donald Atwood, *Memorandum for Secretaries of the Military Departments: Subject: Strengthening Depot Maintenance Activities*, June 30, 1990.

included inter-Service competition for maintenance, competition between Service organizations and private firms, and increased use of depot capacity.[11]

The Atwood directive established a Defense Depot Maintenance Council (DDMC) composed of representatives from the Services and relevant agencies to advise the Assistant Secretary of Defense (Production and Logistics) on maintenance and to coordinate activities. To develop cost-saving strategies, the DDMC commissioned studies on capacity utilization, performance measurement, information systems, cost comparability, and a number of specific weapon systems and technologies. (See table 5-3.)

The current planned changes assume that a major in-Service maintenance base will continue long into the future. The position of DDMC is that

> . . . the highly developed capability of organic maintenance depots, supplemented by that of commercial industry, makes it possible to maintain a high state of readiness during peacetime and sustain the continuing maintenance requirements essential during wartime.''[12]

This position is supported by Public Law 100-370 (July 1988), which directs the DoD to maintain a core logistics capability for performing depot maintenance. The definition and uses of a core capability are discussed later in this chapter.

As a result of the Defense Management Report Decision (DMRD-908) dated November 17, 1990, the Services developed a Corporate Business Plan in December 1991 that describes how the Services will reach the savings goals established earlier by Mr. Atwood. The savings target of $3.9 billion is to be achieved by fiscal year 1995 through increased efficiencies in depot maintenance operations.[13] An initial aim of the Corporate Business Plan appears to be to promote more cost-effective operations while maintaining a depot infrastructure for each Service.[14] The plan is to achieve savings through ''inter-Servicing'' (developing single DoD sites to maintain similar technologies or systems for all Services), increased capacity utilization (consolidating facilities for a given technology and weapon system within each Service), and greater reliance on competition. Current Service depot maintenance structure and restructuring plans are outlined below.

Table 5-3—Defense Depot Maintenance Council Commodity Studies

Army lead
Rotary wing aircraft
Combat, artillery, and tactical wheeled vehicles
Gas turbine engines/compressors
Conventional munitions
Rail equipment
General purpose equipment
Navy lead
Carrier based aircraft[a]
Tactical missiles
F-4 and OV-10 aircraft[a]
Flexible computer integrated manufacturing
Remotely piloted vehicles/unmanned aerial vehicles
J79/T56 engines[b]
Air Force lead
Land based aircraft[a]
Type I metrology laboratories
Landing gear
TF30/F110/LM2500 engines[b]
Engine blades/vanes
Bearings
Marine Corps lead
Small arms
Ground communications-electronics equipment
DLA lead
Industrial plant equipment

[a] Combined into one fixed wing aircraft study.
[b] Combined into one engine study.

SOURCE: Defense Depot Maintenance Council, Corporate Business Plan FY 91-95, December 1991.

Army Depot Maintenance

The Army's depot level maintenance is managed by the Army Materiel Command through its Depot System Command (DESCOM), which administers maintenance funds and assigns work to depots.[15] The Army currently runs 8 major depot maintenance facilities, has a budget of about $1.6 billion, and

[11] Ibid.

[12] Defense Depot Maintenance Council, op. cit. footnote 6, p. 5.

[13] Ibid., p. 1.

[14] Depot Maintenance Executive Group, *Depot Maintenance Business Vision and Strategies for 1995 and Beyond: Findings for the Joint Policy Coordinating Group*, Aug. 21, 1991.

[15] Depot maintenance requirements are determined by The Army Materiel Command's six major subordinate commands: Armament, Munitions and Chemical Command (AMCCOM); the Aviation Systems Command (AVSCOM); the Communications-Electronics Command (CECOM); the Missile Command (MICOM); the Tank-Automotive Command (TACOM); and the Troop Support Command (TROSCOM).

Table 5-4—Army Organic Depot Maintenance Facilities

Facility	Location
Anniston Army Depot	Anniston, Alabama
Corpus Christi Army Depot	Corpus Christi, Texas
Letterkenny Army Depot	Chambersburg, Pennsylvania
Mainz Army Depot[a]	Mainz, West Germany
Red River Army Depot	Texarkana, Texas
Sacramento Army Depot[b]	Sacramento, California
Tobyhanna Army Depot	Scranton, Pennsylvania
Tooele Army Depot	Tooele, Utah

[a] Closing.
[b] Designated for closure by "Defense Base Closure and Realignment Commission Report," July 1, 1991.

SOURCE: Defense Depot Maintenance Council, *Corporate Business Plan FY 91-95*, December 1991.

employs about 18,000 people in its in-Service facilities.[16] (See table 5-4.) In fiscal year 1990 the program repaired over 300,000 secondary items (e.g., radios) and almost 100,000 major end items (e.g., tanks, trucks, engines).[17]

Over the past decade, the Army has contracted out to private firms between 30 and 40 percent of its total depot work. The percentage contracted out varies by type of equipment. For example, in fiscal year 1989, about 50 percent of Army aviation depot maintenance went to private firms, and another 10 to 15 percent was sent to the other Services. In contrast, only about 35 percent of vehicle maintenance was done outside of the organic base, and over 90 percent of that was performed in a government-owned, contractor-operated (GOCO) facility.[18] The amount of maintenance contracting has been controversial. Current legislation requires that not less than 60 percent of funds available for fiscal years 1992 and 1993 Army depot level maintenance shall be used for maintenance performed by employees of the Department of Defense.[19] (Congress' role in legislating different private and public-sector mixes of military maintenance is discussed below.)

As a result of DDMC actions, the Army is engaged in a significant restructuring and consolidation by technology and type of equipment at single sites. (See figure 5-1.) Heavy combat vehicle maintenance will be consolidated at Anniston Army Depot, light combat vehicles and artillery at Red River, missiles at Letterkenny, and tactical vehicles (e.g., trucks) at Tooele. Further steps for achieving savings involve the increased use of inter-Service maintenance, and the closing of both the Sacramento Army Depot and the Mainz Army Depot in Germany.

Photo credit: U.S. Army

Anniston Army Depot performs depot level maintenance on the Army's tanks.

While these steps promise increased peacetime utilization of the remaining facilities, they also carry the risk that depots may be less responsive in crisis or war. Army maintenance planners express concern that excessive consolidations could impair their ability to react to contingencies like Operation Desert Storm. While they acknowledge the important support of private contractors during the Persian Gulf War, they argue that the Army's in-Service capability is essential to support future theater contingencies. Indeed, the Army's maintenance base strategy anticipates that the percentage of future maintenance carried out in government facilities will increase.[20] The Army's flexibility in reducing the percentage of maintenance in government facili-

[16] U.S. Army Depot System Command briefing, Nov. 13, 1991. DESCOM employs more than 30,000 personnel. The remainder of these personnel are involved in meeting the command's other responsibilities: ammunition storage, maintenance of portions of the Nation's strategic materials stockpile, and the distribution of commodities assigned by the Army Material Command, the Defense Logistics Agency, the General Services Administration, and other suppliers.

[17] U.S. Army Depot System Command Director for Maintenance briefing, Sept. 30, 1990.

[18] Kiebler et al., op. cit., footnote 7.

[19] National Defense Authorization Act for Fiscal Years 1992 and 1993, sec. 314.

[20] Defense Depot Maintenance Council, op. cit., footnote 6, pp. 12-13.

326-447 - 92 - 5 : QL 3

Figure 5-1—Army Depot Maintenance Realignment: Work Consolidation

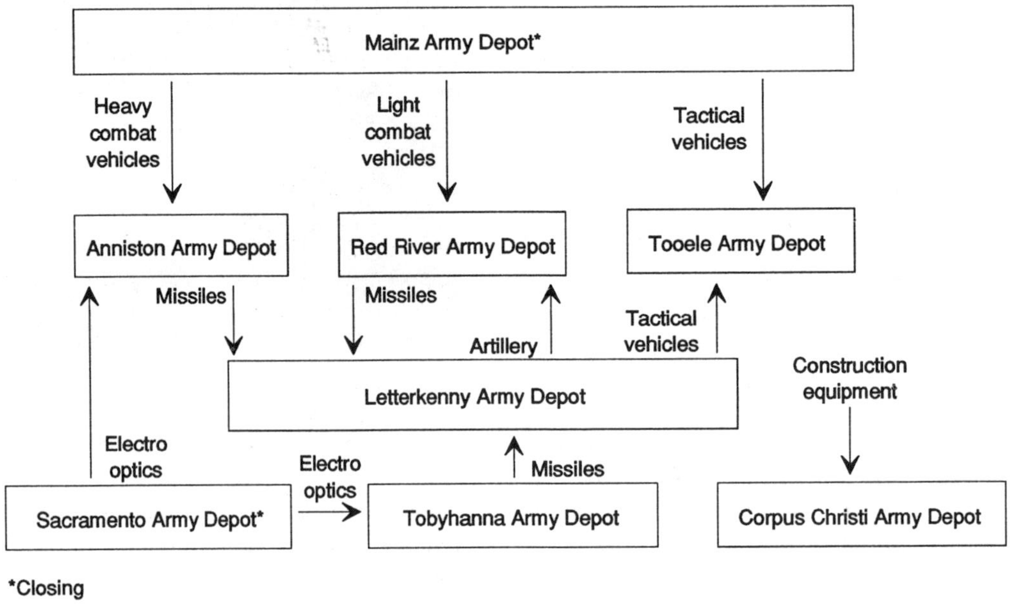

NOTE: This figure does not include inter-Service transfers.
SOURCE: DESCOM command briefing.

ties (should it decide such reductions are best) is limited by legislation.

DESCOM will soon be consolidated with a portion of the Army Armament, Munitions, and Chemical Command into a single Army Industrial Operations Command. This new command is expected to consolidate the depots into smaller, robust manufacturing and maintenance centers that will focus on maintaining those military systems used in the short-warning regional conflicts that Army planners believe are the most likely contingencies in the foreseeable future.

Navy Maintenance and Overhaul

Navy depot maintenance and overhaul is managed by two organizations. The Naval Air Systems Command (NAVAIR) controls the six Naval Aviation Depots. The Naval Sea Systems Command (NAVSEA) manages the public shipyards (table 5-5) and controls the repair and overhaul work conducted at private shipyards. NAVSEA also manages ordnance facilities and weapons stations that perform depot level maintenance. Total Navy maintenance was over $6.5 billion in fiscal year 1991.

Navy ship repair and overhaul is conducted at 44 private shipyards and 8 Navy shipyards. The workforce engaged in Navy repairs and overhaul work consists of about 20,000 in the private sector (out of a total private shipyard workforce of just over 100,000) and 60,000 public-sector workers (which the Navy plans to reduce to about 40,000). Additionally, U.S. Navy ships whose home ports are outside the United States are overhauled overseas. For example, Navy overhaul and repair activities at Subic Bay, Philippines; Guam; and Yokosuka, Japan have, in recent years, totaled more than $100 million per year.

In the mid-1960s, the Navy adopted a policy of assigning all new ship construction to private shipyards and having its own shipyards concentrate on overhaul and repair. Since that time, 60 to 70 percent of the Navy's ship repair and overhaul work has been done by Navy shipyards, while the remaining work, along with all new construction, has been performed by private-sector yards.[21] Congress required competition between the private and

[21] Clinton H. Whitehurst, Jr., *The U.S. Shipbuilding Industry* (Annapolis, MD: Naval Institute Press), p. 90, notes that between 1974-1983 about 30 percent of Navy ship repair went to private shipyards.

Table 5-5—Navy and Marine Corps Depot Maintenance Facilities

Facility	Location
Naval Aviation Depots	
NADEP Alameda	Alameda, California
NADEP Cherry Point	Cherry Point, North Carolina
NADEP Jacksonville	Jacksonville, Florida
NADEP Norfolk	Norfolk, Virginia
NADEP North Island	San Diego, California
NADEP Pensacola	Pensacola, Florida
Naval Shipyards	
Naval Shipyard Charleston	Charleston, South Carolina
Naval Shipyard Long Beach	Long Beach, California
Naval Shipyard Mare Island	San Francisco, California
Naval Shipyard Norfolk	Portsmouth, Virginia
Naval Shipyard Pearl Harbor	Pearl Habor, Hawaii
Naval Shipyard Philadelphia[a]	Philadelphia, Pennsylvania
Naval Shipyard Portsmouth	Portsmouth, New Hampshire
Naval Shipyard Puget Sound	Bremerton, Washington
Ship Repair Facilities	
SRF Guam	Guam, Mariana Islands
SRF Yokosuka	Yokosuka, Japan
Marine Corps Logistics Bases	
MCLB Albany	Albany, Georgia
MCLB Barstow	Barstow, California

[a] Scheduled to be closed.

SOURCE: Defense Depot Maintenance Council, *Corporate Business Plan FY 91-95*, December 1991.

Photo credit: Bath Iron Works

The *U.S.S. Samuel B. Roberts* at the Bath Iron Works undergoing extensive repair for mine damage. Many types of maintenance and repair work require the same skills and facilities as manufacture.

public sectors for Navy ship overhaul and repair beginning in 1985. By 1990, largely because of declines in shipbuilding work, repair constituted almost 20 percent of the work in private yards doing business with the Navy.[22]

The Navy Sea Systems Command's plan for achieving its DMRD-908 savings goals includes consolidations and reductions in workforce that began in fiscal year 1991. The Navy reports that its ship depot level maintenance resides in both private and public yards, although the near total lack of new construction threatens the survival of private U.S. shipyards. The Philadelphia Naval Shipyard is scheduled to close once it completes work on the aircraft carrier *Kennedy*. The Navy has recently proposed consolidation of its ship overhaul capabilities by creating a central-hub shipyard on both the east and west coasts that would provide "support functions such as planning, design, procurement and accounting and maintenance."[23] This consolidation of overhead functions would support a series of satellite yards that retain unique facilities (e.g., nuclear submarine overhaul and repair, and aircraft carrier overhaul) and trained personnel. Despite the consolidations and closures to date, and those that are planned, there is still considerable overcapacity in U.S. shipbuilding and repair. Nevertheless, the Navy's ability to consolidate further may be constrained by the huge capital investments in dry docks and support equipment required in its specialized maintenance facilities.

Naval aviation depot maintenance, employing more than 20,000 people, is carried out in 6 Naval Aviation Depots, which also benefited from the modernization of the 1980s. In response to DMRD-908, the Navy plans by fiscal year 1992 to consolidate maintenance activities for each type of aircraft at single sites.[24] Plans call for the 6 depots to be

[22] Naval Sea Systems Command, *United States Shipbuilding Industry*, briefing paper, July 1990. Navy shipbuilding covered the bulk of all other work. The latest Navy report to Congress on shipbuilding, *Report on the Effects of the FY 1991-97 Navy Shipbuilding and Repair Programs on U.S. Private Shipyards and the Supporting Industrial Base*, April 1991, noted that "In recent years, Navy funding has accounted for 90 percent of the employment at those private yards performing Navy work. Further, 90 percent of Navy shipbuilding funds has been concentrated in only five private yards."

[23] Robert Holzer and Neil Munro, "Navy Weighs Overhaul of Shipyards," *Defense News*, Dec. 23, 1991, p. 17.

[24] Defense Depot Maintenance Council, op. cit., footnote 6, p. 15. The A-6 will not be single-sited until the completion of current rewing work.

Table 5-6—Air Force Organic Depot Maintenance Facilities

Facility	Location
Air Logistics Centers	
Ogden Air Logistics Center	Hill Air Force Base, Utah
Oklahoma City Air Logistics Center	Tinker Air Force Base, Oklahoma
Sacramento Air Logistics Center	McClellan Air Force Base, California
San Antonio Air Logistics Center	Kelly Air Force Base, Texas
Warner Robins Air Logistics Center	Robins Air Force Base, Georgia
Other Air Force depot maintenance activities	
Aerospace Guidance and Metrology Center	Newark Air Force Station, Ohio
Support Group Europe[a]	RAF Kemble, United Kingdom
Detachment 35	Kadina Air Force Base, Japan
Aerospace Maintenance and Regeneration Center	Davis-Monthan Air Force Base, Arizona

[a] Scheduled to close.

SOURCE: Defense Depot Maintenance Council, *Corporate Business Plan FY 91-95,* December 1991.

linked through two Business Operating Centers co-located with the depots at Norfolk, Virginia and North Island, California. According to the Navy, savings will be achieved by reducing the number of personnel who now work on a single aircraft type at more than one site, and through equipment reductions.[25] Engine and aircraft component work is being consolidated, and Navy plans also call for increased inter-Service maintenance. The aviation depots, like the naval shipyards, have engaged in limited competition with commercial firms since 1987. The Navy projects savings from competition in aircraft maintenance to add up to more than $550 million between fiscal year 1992 and fiscal year 1995.

Marine Corps Depot Maintenance

The Marine Corps has two logistics bases. (See table 5-5.) The Service has done little outside contracting and has used Navy facilities to support Marine aviation. Pursuant to DMRD-908, the Marine Corps plans ''cost avoidance'' of about $27 million by not developing its own Abrams tank maintenance facilities. It also anticipates additional savings from increased inter-Service maintenance combined with increased competition. Indeed, most proposed Marine Corps savings are expected to come from increased efficiency resulting from greater competition, both among the Services and between the public and private sectors.

Air Force Depot Maintenance

The depot level maintenance activities of the Air Force are currently managed by the Air Force Logistics Command and include the repair, modification, and support of aircraft and equipment.[26] The Air Force has five major Air Logistics Centers (ALCs), some smaller support centers, and a limited depot maintenance capability overseas. (See table 5-6.) Air Force maintenance currently employs about 36,000 people (scheduled to fall to about 31,000 by 1995). Fiscal year 1991 work totaled almost $4.7 billion. The Air Force performs about 60-70 percent of its depot maintenance in its ALCs.[27] Another 6 percent of Air Force depot work is performed by the other Services,[28] and the remainder is performed by private firms under contract. In fiscal year 1988, the Air Force Logistics Command repaired or modified 1,307 aircraft, 7,727 engines, and 817,000 exchangeable parts. Approximately 90 Air Force systems are currently supported throughout their life by the private sector.

The Air Force modernized its depot maintenance system during the 1980s. It has long consolidated its depot maintenance around Technology Repair Centers. For example, the repair of aircraft engines is concentrated at the Oklahoma City and the San Antonio ALCs and landing gear at Ogden ALC. The Air Force is now downsizing and further consolidat-

[25] Ibid., p. 14.

[26] On July 1, 1992 the Air Force Logistics Command and Air Force Systems Command will merge into the Air Force Materiel Command.

[27] MG Joseph K. Spiers, commander, Oklahoma City Air Logistics Center, testimony before the House Committee on Armed Services, Panel on the Structure of the U.S. Defense Industrial Base, Nov. 1, 1991.

[28] Air Force Logistics Center, *1988 Production Base Analysis,* October 1989, p. *ix.*

ing its base to reduce total costs by about $1.1 billion between fiscal year 1991 and fiscal year 1995.

Current Air Force mobilization planning requirements are based on a scenario that envisions two simultaneous regional conflicts in different parts of the world.[29] Air Force studies of the infrastructure needed to support these requirements caused the Service to begin downsizing the workforce beginning in fiscal year 1991. The Air Force plans to retain, but scale down, each of the current ALCs.

The Air Force plan calls for rapid personnel reductions, installation closures, and process improvements. In the longer term, savings will be accomplished by increased inter-Service maintenance competition and increased utilization of facilities. Most long-range savings are expected to come from greater efficiency spurred by competition.[30]

The Air Force considers the retention of skilled personnel an immediate and important problem and is concerned about their loss as budgets and forces decline. The commander of the Oklahoma City ALC, for example, testified that he was losing both current expertise and future capability because his older, experienced workers were leaving through "early-out" retirements and his younger workers were leaving because of reductions-in-force.[31] A second Air Force concern is its ability to continue sufficient investment in depot facilities over the long term.

REQUIREMENTS FOR A ROBUST FUTURE MAINTENANCE BASE

Congress and the DoD need to plan the size and nature of the future maintenance base. As discussed at the beginning of this chapter, the future base will be smaller because the Nation will have fewer deployed weapons, will face smaller military threats, and systems in the field may be more reliable. Even so, the retention of older weapon systems will make maintenance, as well as upgrading and retrofitting, important. Much of this upgrading is expected to occur in avionics, electronics, software, and advanced materials. Thus, a future robust maintenance base will not be just a collection

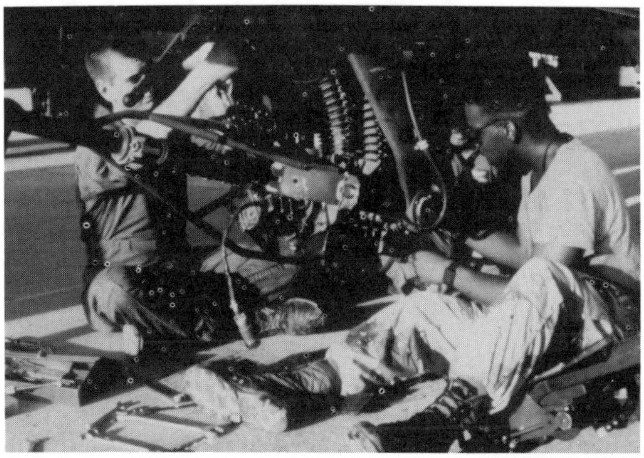

Photo credit: DoD

In modern combat, field maintenance must be able to return equipment rapidly to operation.

of metal-working shops but must be capable of supporting an increasingly complex inventory of weapon systems.

The future maintenance base must be efficient in peacetime; this is an objective of many current Service initiatives. The Services' plans for future efficiency rest on the increased use of competition and better use of physical plant. But competition and high facility utilization can be incompatible. True competitive bidding implies multiple sources, and hence some overcapacity. The anticipated savings through competition hoped for by each Service may be based on the belief that it, and not another Service, will win such competitions. A major bonus of increased competition is that it is a politically acceptable way of eliminating facilities (public as well as private) that are unable to modernize adequately. Another way to improve efficiency is through new maintenance techniques and technologies, including modular repair centers, robotics, and advanced diagnostic equipment. Built-in diagnostics may reduce field maintenance costs in the future.

The future base must retain a capacity to respond rapidly to crisis or war. However, peacetime efficiency resulting from the high utilization of the maintenance base in peace may also be incompatible with responsiveness in crisis and war. Such responsiveness will continue to be critical in the short-warning regional conflicts that many planners envi-

[29] Defense Depot Maintenance Council, op. cit., footnote 6, pp. 34-35.

[30] Ibid.

[31] Spiers, op. cit., footnote 27.

sion in the future. During the recent Gulf War, the depot system appeared to respond well. The Services modified equipment both in the United States and in the Gulf. The Army DESCOM, for example, shipped 500,000 tons of materiel, rapidly upgraded 743 M1A1 Abrams tanks, deployed 2,000 civilian employees to the theater, and established a forward maintenance facility in Saudi Arabia. Contractors also deployed hundreds of maintenance personnel to the theater of operations and made important modifications to equipment once they were there. Although the future maintenance base will need to respond, there will be a lesser magnitude of wartime demand associated with the smaller contingencies likely in the foreseeable future.

Photo credit: U.S. Navy

A nuclear submarine leaves the dry dock at Bangor, Washington. Some specialized maintenance work, such as repair and overhaul of nuclear-powered submarines, can only be accomplished at special sites.

ALTERNATIVES FOR THE FUTURE

The principal alternatives for ensuring a robust depot level maintenance base in the future are evident in the Service's responses to DMRD-908, and were noted earlier in the chapter. They are discussed below and include:

1. reduction and consolidation of current in-Service and private-sector capabilities;
2. increased use of the private sector both to gain expected efficiencies, and also to provide support for the private production base;
3. increased use of competition;
4. development of new technology for maintenance; and
5. maintenance and upgrading of U.S.-produced equipment abroad as well as foreign-produced equipment.

These alternatives, as noted earlier, are not mutually exclusive; rather in combination they could fashion a future robust maintenance base.

Reduction and Consolidation

Reduction and consolidation of the base is ongoing, as described above. Current plans are significant in the context of past DoD attempts at consolidation. But viewed in the context of the end of the cold war, these plans are less impressive. Even after present plans are carried out, the DoD will have almost the same number of major in-Service maintenance facilities as existed during the cold war.

Further consolidation can be carried out across Service lines. Consolidating maintenance of similar systems or technologies at single facilities—regardless of Service affiliation—reduces overhead and makes better use of specialized capabilities. Projected DMRD-908 savings from inter-Service maintenance are $120 million for the fiscal year 1991 to 1995 period. The Services report that over 60 percent of depot work could be accomplished by more than one Service. This figure excludes work that requires such specialized facilities as large drydocks, large hangers, naval nuclear-reactor refueling facilities, and the skilled people to run them. Nevertheless, in fiscal year 1989 only 6 percent of the maintenance that could be performed on an inter-Service basis was sent across Service lines, indicating considerable redundancy in the base. That percentage is projected to rise to only about 9 percent by fiscal year 1995.[32]

Individual Service planners express a number of concerns about inter-Service maintenance consolidation. One of the principal worries is that another Service will not meet the special requirements of particular equipment, such as the Navy's need to protect its aircraft engines from the corrosive effects of the marine environment. Other risks cited are possible lack of responsiveness by another Service.

The ongoing reductions and consolidations are politically unpopular because they carry up-front costs that may seem large compared to the promise

[32] Defense Depot Maintenance Council, op. cit., footnote 6, pp. 37-44.

of future, long-term savings. In particular, Service depot maintenance facilities are important sources of jobs. Public opposition to impending closings has led Congress to mandate work assignments to particular facilities to keep them open. But congressionally mandated workloads make it difficult for the DoD to improve the efficiency of the maintenance base. A Logistics Management Institute (LMI) study found, for example, that special legislation enacted in 1986 exempting the Army electronics repair depots from personnel reductions resulted in significant inefficiencies in managing depot workload.[33] Thus, even when Services decide to consolidate facilities, they may be barred from doing so because of congressional pressure to preserve jobs.

The expected large reductions in maintenance requirements, combined with falling defense budgets, make it imperative to rationalize the depot maintenance base. One way of dealing with this problem is legislation such as the Defense Base Closure and Realignment Act of 1990, which requires that all installations be considered equally for possible realignment or closure. The Act established new procedures for closing military installations in the United States and formed an independent commission to recommend which bases should be eliminated.[34] Such legislation can help assert the *national* interest in rationalizing the maintenance base over the *local* interest in preserving jobs.

Changing the Private/Public Mix

The current debate over increasing the percentage of private-sector involvement in future depot maintenance work is principally motivated by two factors: (1) the anticipated reduction in new weapons production that will leave large defense manufacturing firms with little new production for the foreseeable future; and (2) the reduction in spending in the depot base that is driving the consolidations and reductions discussed above. Advocates of more private-sector involvement argue that the private sector can provide depot maintenance at lower cost than can the public sector, and that a shift toward the private sector would help keep the production base healthy during a period of much reduced new weapon procurement.

There are, however, concerns about the long-term implications of increasing the private-sector share of depot maintenance and skepticism about the utility of using depot maintenance to support manufacturing skills. The concerns center on questions about how well the private sector can respond to short-notice crisis and conflict requirements, and whether private contractors can indeed provide depot maintenance at a lower cost. The skepticism centers on the amount of overlap between maintenance and manufacturing skills and whether performing maintenance can indeed support relevant manufacturing skills. It is worth noting, however, that most allies in Europe and the Far East rely on their private sectors for almost all their military depot maintenance.

The Current Mix

A significant portion of depot maintenance funding is currently spent in the private sector. For example, between 20 and 30 percent of the depot level maintenance is now performed by private firms. Almost all new weapon systems begin their service lives under interim contractor support (ICS) provided by the manufacturer of a system. This support usually lasts until the system is deployed in sufficient numbers to warrant transferring maintenance responsibility to the Services. During this initial period, test equipment is developed for use in the Service support base, and Service depot personnel receive maintenance training. While the majority of systems move on to Service depot maintenance, some continue to be maintained by the private sector in what is termed contractor logistics support (CLS).

In addition to the direct revenue from maintenance, repair, and overhaul, the private-sector production base also derives considerable income from the sale of spare parts and other goods and services. The commander of the Air Force Oklahoma City ALC, for example, reported that in fiscal year 1991, his command had contracted for "$2 billion of work with over 6,500 private sector organizations in 46 states and 9 foreign countries."[35]

[33] Kiebler et al., op. cit., footnote 7, p. 2-23.

[34] General Accounting Office, *Military Bases: Observation on the Analyses Supporting Proposed Closures and Realignments*, GAO/NSIAD-91-224, May 1991, p. 15.

[35] Spiers, op. cit., footnote 27. Note, however, that much of this is accounted for by the purchase of spare parts to support work actually done at the ALC.

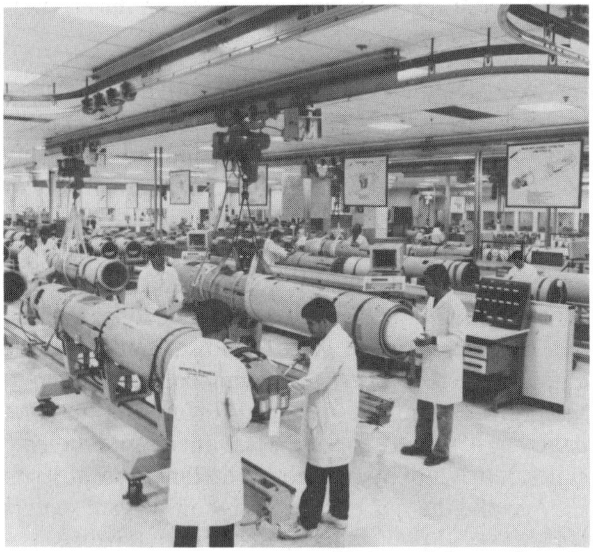

Photo credit: General Dynamics

General Dynamics performs periodic depot maintenance on Navy fleet missiles.

The combination of private-sector maintenance and direct sales accounts for more than 50 percent of current depot maintenance spending. This share demonstrates a private-sector commitment to depot level maintenance and an acceptance by the Services of private-sector involvement. It also suggests the limits of any additional private-sector shift.

The Logistics Core

Current Service and DoD policy of maintaining a "core" in-Service logistics capability will affect any shift to the private sector. The core depot maintenance capability is basically that minimal combination of people and facilities each Service believes it needs to support its forces in likely future operations. According to the Defense Depot Maintenance Council, the logistics core is "an integral part of a depot maintenance skill and resource base which shall be maintained within depot activities to meet contingency requirements."[36] It is to consist of only a "minimum level of mission essential capability."

How this concept of a core capability is determined differs among the Services. The Army, for example, defines its core requirements as workloads that are essential to the mission or critical to the capability of each unit. Navy aviation core requirements are based on a regional war scenario. The Navy's core maintenance requirements for its sea forces are defined as "a responsive, geographically dispersed, strike-free industrial capacity . . . . whose priorities are controlled by the Navy." Interestingly, the Navy's logistics core for sea systems includes private as well as Service facilities and people. The Air Force definition of core requirements is based on an analysis of the skills and weapons needed to support specific regional-conflict scenarios.[37]

Commercial firms will have a difficult time competing with in-Service depots for future maintenance work if the Services reserve a large core for themselves. While the Services' protected logistics core can reduce their own workload fluctuation and maintain internal skills, it has the drawback of increasing the fluctuation in any workloads performed in the private sector. From the Services' perspective, however, the concept of a core capability is critical to maintaining essential expertise. They believe that opportunities for changing the private/public mix of maintenance work will be limited because, as one Air Force commander testified, "government workloads that would be the most attractive to the commercial repair and maintenance sector would be the high-volume, state-of-the-art technology, stable workloads . . . [that are] the very workloads that are imperative for [the Air Force] to keep . . . to maintain a mobilization skills base."[38]

The Debate over the Mix

The past division of labor demonstrates industry's interest in depot maintenance. Much of the increased interest is in upgrading currently deployed systems. As part of an integrated DTIB strategy, shifting this work to private firms could provide employment for production staffs during low points in procurement cycles and could also generate another source of income for firms attempting to maintain research and design, as well as production, capabilities. Proposals for upgrading armored vehicles, for example, envision work-sharing arrangements with Service depots to combine depot overhaul with major upgrades and thus keep production lines warm. Upgrading could bring older tanks and infantry fighting vehicles up to date with new communica-

[36] Defense Depot Maintenance Council, op. cit., footnote 6, p. 33.

[37] Ibid., pp. 33-35.

[38] Spiers, op. cit., footnote 27.

tion and sensor technologies.[39] Upgrades would also support the production base through the manufacture of subcomponents and parts. In fact, upgrades will have a positive impact on subtier firms regardless of whether the overall system work is performed by a prime contractor or an in-Service facility. The Electronic Industries Association (EIA), for example, anticipates that upgrades will provide considerable business for the electronics sector in the next decade, although recently, EIA's estimates of the size of the future market have been going down.[40]

In general, manufacturing firms argue that they have an inherent capability to maintain the equipment they have produced and that developing in-Service support capabilities often "duplicates an existing commercial defense capability that was developed to design and initially manufacture the military equipment. As such, it is entirely feasible in many cases for the U.S. [private] industrial base to replace the in-Service capability for the U.S. military."[41] Contractors say they are more efficient because the pressures of competitive bidding force them to control costs and that they have different personnel policies than does the DoD. Further, many private contractors say that they are as responsive as the Service maintenance base and point out that current Service response capabilities already depend on spare parts and services from private industry. The private sector's ability to respond quickly has been demonstrated during the Vietnam and Persian Gulf conflicts.

Government proponents of an in-Service capability make a number of counterarguments. They believe that in-Service maintenance facilities are more responsive in crisis and war than private industry, and are also more flexible because they can take on new work without changes in contracts. The Air Force, for example, has testified that the Air Logistics Centers have "the flexibility to deal with a highly dynamic war environment and that no contractor or group of contractors could replace the cohesive, highly flexible capabilities of the in-Service facilities."[42] Along similar lines, the Army argues that assigning surge maintenance tasks "to the private sector, without the insurance of the contractor's ability to rapidly expand, could jeopardize the Army's ability to get equipment to the soldier in time of national emergency."[43] The Navy has expressed less concern about increasing private-sector involvement in maintenance than the other Services. In part this is because shipyards are large and easy to monitor and also expensive to duplicate. The Navy plans for private shipyards to be the sole provider of some of its ship maintenance and considers private-sector yards to be part of its core sea systems capability.

The Services have noted that many systems maintained in contractor logistics support in the past have ultimately devolved to in-Service maintenance as they aged and became more difficult to repair. As a result, Service officials are concerned that they will be stuck with maintaining all the old systems rather than those essential to war-fighting. Other risks associated with relying on the private sector are said to be strikes and bankruptcies. A Logistics Management Institute study concluded, however, that these problems are likely to occur only in peacetime and can be dealt with by the DoD through existing legal mechanisms.[44]

The evidence supporting arguments on either side is scarce and largely anecdotal. Some General Accounting Office (GAO) studies have questioned the economics of developing in-Service depot support capabilities for equipment that may be widely used commercially. A recent GAO study, for example, found that the Air Force had spent millions of dollars establishing a maintenance capability for the new engine of the KC-135 tanker but was using only about 15 percent of that capacity. GAO argued that the Service might better have relied on existing

[39] Caleb Baker, "Army Seeks Stable Bradley Production," *Defense News*, Oct. 14, 1991, p. 8.

[40] Breck W. Henderson, "Stagnant Military Electronics Spending Likely Under Tight 1990s Budgets," *Aviation Week and Space Technology*, Mar. 16, 1992.

[41] Mr. Gordon R. England, Executive VP-Aircraft Programs, General Manager-Fort Worth Division, General Dynamics, *Statement before the House Armed Services Committee, Structure of the U.S. Defense Industrial Base Panel*, Oklahoma City Field Hearing, Nov. 1, 1991.

[42] Spiers, op. cit., footnote 27.

[43] Army Information Paper, *Army's Maintenance and Logistics System*, May 14, 1991, provided in response to OTA questions.

[44] Kiebler, et al., op. cit., footnote 7, pp. 2-24 to 2-26.

326-447 - 92 - 6 : QL 3

commercial facilities.[45] While a GAO study on the effects of competition found that using private/public competition has resulted in savings in naval aviation overhaul, an earlier study of ship repair could not confirm the savings the Navy had projected from greater use of the private sector.[46] Discussions with government personnel indicate a belief that private firms are less responsive in peacetime (because of general business practices), but OTA has been unable to find any hard evidence to show that the Service maintenance base is indeed more responsive than private contractors in a crisis.

Despite arguments that private contractors avoid maintaining older equipment, the Army has contracted out maintenance of older electronics equipment that its own depots do not want to handle. Any migration of older systems into the in-Service base may stop as the Services are forced to pay the true costs of maintenance whether it is performed in the in-Service component of the base or in the commercial component.

Increased maintenance, especially overhaul and upgrades, may help support the production base in some sectors. Upgrades of several armored vehicles might maintain active production lines. Further, some sectors (such as electronics) claim there is considerable overlap in skills between maintenance and manufacturing. Nevertheless, many government planners remain skeptical of the overall benefits of such change. They believe that industry is more interested in production than maintenance and is therefore unreliable, that maintenance skills are different from manufacturing skills, and that DoD efforts to support production will reduce Service maintenance capacity while propping up uneconomical production. Further, the use of private firms could erode surge capability over time. The basis for many of these government concerns is best summarized in an observation by Air Force logistics planners:

> Transferring maintenance tasks to the private sector will provide short term capital to defense firms. Over time, however, it is likely that private sector firms will evolve to "peacetime efficient" operations with little of the "excess capacity" needed for the essential support of any significant surge. We will have canceled the insurance policy (i.e., organic capability) in anticipation of only "good times." If the "good times" end quickly we will be at a significant logistics disadvantage.[47]

Photo credit: Bath Iron Works

This plasma arc burning machine cuts metal pieces for both new construction and the maintenance and overhaul of older ships.

Sorting through the arguments on both sides demands systematic study.

Congress

Congress has exhibited a mixed response to increasing private-sector involvement in depot maintenance. For example, Congress and the Navy have sought to ensure that the private-sector share of ship repair not fall below 30 percent.[48] At the same time, Congress has limited private-sector involvement in Army and Air Force depot maintenance to not more

[45] U.S. General Accounting Office, *Commercial Practices: Opportunities Exist to Reduce Aircraft Engine Support Costs*, GAO/NSIAD-91-240, June 1991, p. 5. The Air Force spent $40 million on a repair facility at the Oklahoma City ALC and also opened three intermediate maintenance facilities for the engine. GAO pointed out that General Electric and Aviall repair similar engines.

[46] GAO/NSIAD-92-43, forthcoming, title not available; and U.S. General Accounting Office, *Navy Maintenance: Status of Public and Private Competition*, GAO/NSIAD-90-161, September 1990.

[47] Correspondence with Headquarters, Air Force Logistics Command, Mar. 16, 1992.

[48] Whitehurst, op. cit. footnote 21, p. 67.

than 40 percent of the funded work.[49] Current law supports some public-sector capability. It states that:

> It is essential for the national defense that Department of Defense activities maintain a logistics capability (including personnel, equipment, and facilities) to ensure a ready controlled source of technical competence and resources necessary to ensure effective and timely response to mobilization, national defense contingency situations, and other emergency requirements.[50]

The law gives the Secretary of Defense the authority to identify those in-Service maintenance capabilities necessary to maintain responsiveness. As a result, the Secretary has discretion over when to use the private sector.

Congress may wish to support significant change in the current private/public mix. Should more private-sector involvement appear desirable, there are a number of ways to move in that direction. One is through increased contracting and competition with the private sector, as discussed in the following section. Another way would be to privatize existing in-Service facilities. Depots might be converted to GOCOs. Such facilities have the advantage of combining long-term government control of the facilities with more flexible private-sector operation on a day-to-day basis. If depots become GOCOs, large capital cost items (e.g., dry docks) could be paid for by the DoD while management and workers could be paid through the private sector. As for responsiveness, many ammunition facilities are GOCOs and are designed to respond to crisis. Depots could also be closed and equipment mothballed. While this approach would not address the need for responsiveness, it would provide some capability for longer term mobilization.

Finally, depots could be sold to the private sector. Selling depots to industry may be more difficult than converting them to GOCOs. Privatization could allow the use of the large government investment for commercial as well as military use. However, industry has shown little interest in buying the defense industrial base facilities that have been for sale over the past decade. Industry would probably prefer to move maintenance activities to existing private-sector facilities.

As Service maintenance organizations become more streamlined, the greater efficiency of contractors may become a less compelling argument for moving to the private sector. For example, the Air Force has reduced its workforce by 6,000 since fiscal year 1991 and plans to support many programs with personnel hired on temporary appointments. Conversely, if the United States chooses to move toward more civil-military integration in weapons design and manufacture, increased use of the private sector for maintenance might make even greater sense.

Competition and Efficiency in Military Maintenance

Competition in the maintenance base, like that in the production base, is intended to promote efficiency and fairness. In the past, individual DoD program managers had the authority to decide on the basis of cost whether to rely on in-Service or private maintenance, although the Services planned to maintain some core capabilities. But past policies also stressed the importance of multiple sources for wartime expansion. Thus, these policies often aimed at increasing capacity rather than promoting efficiency.

Competition was also used to help private companies gain access to maintenance contracts. Congressional concern about the health of the U.S. shipbuilding industry resulted in opening Navy ship repair work to private shipyards. The first such competition occurred in fiscal year 1985. By the end of fiscal year 1989, maintenance work on 43 surface ships and 25 submarines had been competed.[51] Competition involving the Naval Aviation Depots began in 1987. The National Defense Authorization Act, passed in fiscal year 1991, expanded the maintenance competition programs on a limited basis to all the Services. This program is designed to promote competition among the Services as well as between the Services and private industry.

DMRD-908 proposes to achieve one-third of the projected $3.9 billion maintenance base savings by

[49] Both the 1988 and 1989 Defense Authorization Acts contained a requirement that the Army spend a minimum of 60 percent of the depot maintenance budget on programs performed by the organic DoD workforce. Current law mandates that 60 percent (by cost) of Army and Air Force depot work be performed by government employees.

[50] 10 U.S. Code Section 2464.

[51] GAO/NSIAD-90-161, op. cit., footnote 46.

efficiency improvements resulting from competition. Competition is expected to increase efficiency, control and reduce costs, and foster innovative approaches to maintenance. Four areas are to be opened to competition:

1. items currently under commercial contract whose renewal is imminent,
2. major refurbishment and modification programs,
3. manufacturing and fabrication, and
4. in-Service workloads deemed in excess of the logistics core.

How effective competition really is in controlling maintenance costs is debatable. As noted above, a 1990 General Accounting Office report on shipyard competition concluded that the Navy's projected cost savings for private yard ship repair could not be substantiated.[52] GAO noted, however, that competition had encouraged the Navy's own shipyards to adopt "a more businesslike approach to ship repair work." In addition, a more recent GAO study of naval aircraft maintenance found that competition in F-14 aircraft maintenance had resulted in a 25 percent decline in overhaul costs.[53]

Partly because of congressional restrictions, competition is just beginning in the other Services. U.S. Code, Title 10, Section 2466, for example, prohibited the Army and the Air Force from competing against each other or with the private sector. Competition will take some time to develop properly as organizations that have never had to compete learn to price their services and put together bid proposals. But the advent of business offices at depots, and new awareness of overall costs, support the Navy's contention that competition reduces overall maintenance costs.

Structuring competition and developing a "level playing field" agreed to by both the private sector and the public sector will probably remain contentious. A key issue has been how to compare costs among different Service depots and between the private and public sectors. The Services jointly developed and published a *Cost Comparability Handbook* to help make these comparisons and eliminate differences in accounting procedures used by various public and private competitors.

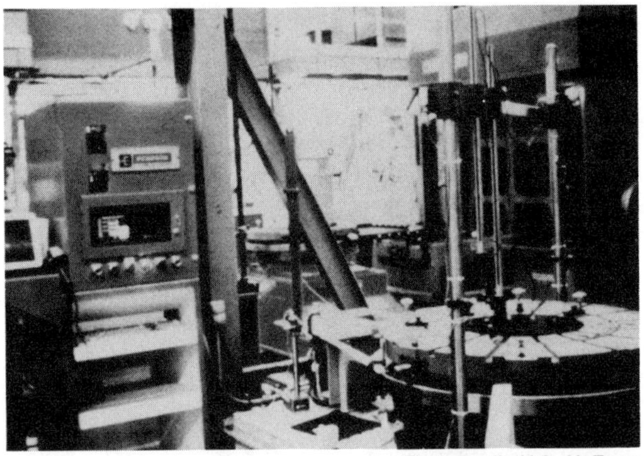

Photo credit: U.S. Air Force

An Air Force repair facility for jet engine casings. This and other repair technologies have been developed with special Repair Technology, or REPTECH, funding.

While recent changes promote competition in the maintenance base, there are still major limitations. Current law limits the Army and Air Force competition program to not "more than 10 percent of all depot-level maintenance of materiel that is not required to be performed by employees of the Department of Defense." This limitation effectively excludes 96 percent of Army and Air Force maintenance work from the pilot program.

Competition, if it develops, may prove to be a good means of selecting those organizations, private or public, that should be retained in the future maintenance base. It will be much more difficult to preserve a government facility or private firm that has systematically failed to attract work on a competitive basis.

New Technology

The future depot maintenance base should seek to benefit as much as possible from new technology. An obvious area for improvement is the design and development of weapon systems and equipment with higher overall reliability, thereby reducing maintenance requirements. Modular components (e.g., circuit boards) and built-in diagnostic checks are changing maintenance tasks. They are, for example, making it easier to repair and replace equipment in the field.

[52] Ibid., p. 1.

[53] GAO/NSIAD-92-43, op.cit., footnote 46.

Flexible manufacturing systems, robotics, and computer-integrated manufacturing are all increasingly used in weapon system maintenance and hold the promise of reducing labor requirements. The Air Force has an active Repair Technology program (REPTECH) as a part of its Manufacturing Technology program. The Service's REPTECH initiatives include a flexible center to repair aircraft engine casings at the Oklahoma City ALC, composite engine repair centers at Oklahoma City and San Antonio, and nondestructive means of inspecting solder joints in printed wiring assemblies. The Navy has developed a Rapid Acquisition of Manufactured Parts (RAMP) project at the Charleston naval shipyard to shorten the time needed to produce spare parts, which can take weeks to obtain from the private sector. The Defense Logistics Agency and the Federal Emergency Management Agency are funding a transfer of a prototype of the RAMP technology to a small manufacturer.

Supporting Military Equipment Abroad

Upgrades of U.S. weapon systems abroad, or foreign systems, are another way to support the U.S. defense maintenance base. The potential market is significant. The upgrade of F-5 fighters, for example, is estimated to be a $3 to $5 billion business in Taiwan and Singapore.[54] Upgrading the F-16A/B, which is in foreign nations' air forces, could be worth another $2 billion.[55]

Upgrades or repairs are not the only options for international activities. In the past, U.S. firms have contracted to establish and run military maintenance organizations and facilities for selected countries (e.g., Iran under the Shah). Maintenance support of allied forces is a possible source of future income.

The U.S. government is involved in several international cooperative maintenance programs through the NATO Maintenance and Supply Agency (NAMSA), the primary logistics support agency for NATO.[56] Since 1985, the United States has increased cooperation with its NATO allies for spare-

Photo credit: DoD

Belgium's is only one of several foreign air forces that fly the U.S. F-16 fighter aircraft.

parts support and depot level maintenance. The United States is involved in collaborative maintenance on the Multiple Launch Rocket System, the PATRIOT Missile System, and the C-130 Hercules aircraft.[57] There are 11 other NATO maintenance partnership programs in which the United States does not participate.

Opportunities for supplying foreign markets with upgrades, or for providing other services, will depend on U.S. technology-transfer policy—as do initial sales of weapons. The tasks for which U.S. firms might be most competitive (e.g., avionics and electronics) might also present the greatest risk for giving away technological and military advantage.[58] There is also likely to be more international competition as foreign firms vie with U.S. firms for the global maintenance market.

SUMMARY

Maintenance is critical to peacetime operations and to sustaining forces in crisis or war. The requirements for depot maintenance have significantly changed as a result of the waning direct military threat to Western Europe. While current DoD efforts to streamline and consolidate the base

[54] William B. Scott, op. cit., footnote 5.

[55] Michael Mecham, "European Partners Pair Upgrades with New Aircraft Development," *Aviation Week & Space Technology*, July 12, 1991, p. 54.

[56] It should be noted that logistics has been a national responsibility in NATO.

[57] Department of Defense, *Combined Annual Report to Congress on Standardization of Equipment with NATO Members of Cooperative Research and Development Projects with Allied Countries*, July 1991.

[58] At the same time, however, supplying maintenance and upgrades may provide leverage on a client. If the maintenance support is cutoff, the weapon system will degrade.

represent significant change by the standards of the past 40 years, they are insufficient in the new security environment. A smaller national DTIB demands that the Nation consider significant changes in the maintenance base.

Consolidation of in-Service maintenance facilities will be constrained by the fact that such facilities are important sources of jobs, and sometimes the largest employer in a region. As a result, there is often considerable political pressure to maintain these facilities.

Responsiveness of the maintenance base in crisis and war will remain important. However, potential regional threats do not demand the magnitude of surge maintenance required in the past. Future maintenance capabilities might therefore stress peacetime efficiency, which could be enhanced by investments in process technology. The Air Force is using its REPTECH Program for such improvements, but the other Services have made more limited efforts in this area. Congress might wish to consider how best to apply new technology to maintenance.

The arguments for transferring more maintenance responsibilities into the private sector include lower costs, less redundancy, and better support of an integrated DTIB. Congress should examine the arguments for increased use of the private sector for maintenance and consider how best to modify the public/private split, for example by transferring maintenance work to private firms or by converting public facilities to GOCOs. Increased competition among in-Service facilities, and between the private and public sectors, may be the best way to accomplish this transition. Such competition could select the facilities best qualified to support future forces over the longer term. Finally, Congress should reevaluate the concept of a core logistics capability now used to define which activities should be retained in the in-Service maintenance base.

Maintenance contracts directed towards critical manufacturers in the private sector may help support the firms in a period of declining defense procurement. But the degree of support will probably vary by industrial sector. Combined with a prototyping-plus strategy that provides for some manufacturing, as well as continued technological innovation, private-sector maintenance might add significantly to the health of the future U.S. DTIB.

Chapter 6

Good, Integrated Management

Contents

Box

Figures

Table

Chapter 6
Good, Integrated Management

INTRODUCTION

Previous chapters examined the three principal elements (R&D, production, and maintenance) of the defense technology and industrial base (DTIB), the desirable characteristics of the future base, and ways to achieve those characteristics. *Redesigning Defense* noted that **good, integrated management is a desirable characteristic of the future DTIB, indeed, it is fundamental to its health and strength.** The report defined integrated management as linking the DTIB goals of crisis and wartime response with the peacetime goals of development and production of high-quality and affordable military materiel. Good management will also closely integrate the R&D, production, and maintenance activities of the DTIB. Thus, there is a need for coordination and cooperation throughout the DTIB.

DTIB management has been the focus of a number of recent studies. One of the most influential was the President's Blue Ribbon Commission on Defense Management (known as the Packard Commission), which recommended a number of reforms. Many of these have been adopted, including the establishment of an Office of the Under Secretary of Defense for Acquisition (USD(A)) responsible for Department of Defense (DoD) acquisition policy, administration, oversight, and supervision; the vesting of similar acquisition authority and responsibility in a single Service Acquisition Executive (SAE) within each Service; and a general simplifying of the DoD management structure.[1] (See figure 6-1.)

The most recent annual report of the Secretary of Defense details additional actions taken by the DoD to improve DTIB management and some successes to date.[2] Despite this attention, there is widespread agreement that management of the future DTIB requires additional changes to meet the challenges of the new world security environment.[3]

Figure 6-1—Acquisition Management Structure

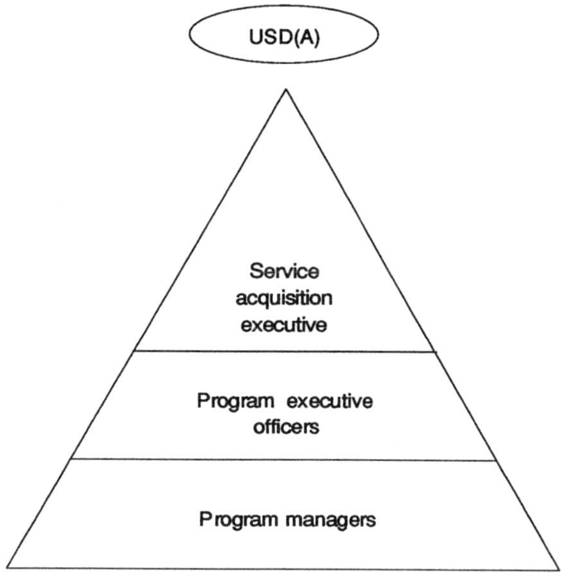

SOURCE: Defense Systems Management College, 1989; Office of Technology Assessment, 1992.

This chapter does not assess all recent management initiatives, but focuses on the management implications of alternatives for restructuring the DTIB discussed in earlier chapters. The chief DTIB management challenge is how to preserve an advantage in defense technology and retain the ability to manufacture and maintain military systems—all on a much smaller defense budget.

Additional management changes are needed to promote integration with the Nation's civilian industrial base and to implement a prototyping strategy as the administration and others, including OTA, have suggested. While the Packard Commission recommended the use of prototyping to test new ideas and to lower costs, its recommendations were advanced in the context of large production runs. As outlined in Chapters 3 and 4, the management of prototyping

[1] The President's Blue Ribbon Commission on Defense Management, *A Formula for Action: A Report to the President on Defense Acquisition*, April 1986.

[2] Secretary of Defense Dick Cheney, *Annual Report to the President and the Congress* (Washington, DC: U.S. Government Printing Office, February 1992), pp. 28-39 and 49-53.

[3] Secretary of Defense Dick Cheney, *Defense Management Report to the President*, July 1989, p. 8. This report noted that "Efforts to date have not produced the tangible results envisioned by the Commission. This is indicative of the dimension of the problems the Commission identified, the far-reaching solutions it offered, and the persistence required if DoD's management of major acquisition programs is to emulate the characteristics of the most successful commercial and government projects."

Table 6-1—Characteristics of Successful Acquisition Programs[a]

- Clear command channels
- Program stability
- Limited reporting requirements
- Small, high-quality staffs
- Good communications with users
- Prototyping and testing

[a] The President's Blue Ribbon Commission on Defense Management, April 1986.

will be affected by a transformed security environment that will feature production of fewer new weapons and longer intervals between new system starts. This new environment requires the elimination of the near-automatic link between program start and quantity production.

The changes in the security environment are likely to require significant shifts in management structure and approach. **Successfully managing a smaller future DTIB will require a much more integrated approach by the DoD, the administration, Congress, and the private sector.** For example, a DTIB that is much more integrated with the Nation's broader industrial base will require managers capable of monitoring civilian technology worldwide.

This chapter examines what good, integrated management of the future DTIB will entail, and considers ways to achieve it in the new national security environment.

GOOD, INTEGRATED MANAGEMENT

Ultimately, the criteria for judging the success of DTIB management will be how well the defense base can 1) provide and support high-quality military equipment at an affordable cost in peacetime; and 2) meet increased military requirements for goods and services in crisis or war.[4] Discussing problems of defense acquisition in the 1980s (many of which still exist), the Packard Commission identified what it felt were certain characteristics common to successful commercial and government projects (table 6-1). The Commission recommended that the executive branch and Congress change DoD acquisition to develop these characteristics. The Commission suggestions still apply and will strengthen the future base, but alone do not address the changes in DTIB requirements brought on by the ending of the cold war.

People

Trained and experienced people will be critical not just at the top but at every level of future DTIB management. If the future DTIB has greater civil-military integration and less stringent military specifications, the need for individual expertise and judgment will increase at each level. The Packard Commission noted, however, that "recruiting the most capable executives for jobs of such importance to the Nation is extremely difficult . . . in the face of current disincentives to entering public service."[5] These disincentives include relatively low pay for senior government managers; but according to many observers, legislation that severely limits post-government employment is an even greater disincentive.

Organization

The overall balance of management activity should change as managers become more concerned about maintaining the base instead of procuring particular systems. The future DTIB will require shifts away from the present focus on weapon systems production to a focus on R&D and prototyping that might provide more opportunities for testing new ideas and alternative ways of performing a mission. DTIB management might use technologies or mission requirements as an organizing management principle in addition to, or instead of, production. An important management principle remains: **the organization must stay small enough to avoid stifling, bureaucratic intrusion and retain flexibility, but large enough to manage the DTIB.** Indeed, some observers of the DTIB say that the impediments of government are less a function of the number of regulations than of the number of regulators.

[4] The Department of Defense has recently outlined similar criteria for DTIB management. In its *Report to Congress on the Defense Industrial Base*, November 1991, p. ES-7, the Department noted that:

> In order to ensure that industry remains capable of producing the weapon systems the Department of Defense requires, the Department of Defense must manage a range of industrial base activities: the peacetime business of equipping and supporting military forces; ensuring industrial preparedness to deal with potential regional contingencies and conflicts; and planning for the reconstitution or expansion of military forces in response to a potential future, global threat.

[5] The President's Blue Ribbon Commission on Defense Management, op. cit., footnote 1, p. 27.

Streamlining and reducing the DTIB management outside the DoD is important too. Secretary Cheney in his Defense Management Report (DMR) noted that limiting any reorganization to the DoD would not be sufficient to truly improve DTIB management. He argued that the base also suffers from the way Congress carries out its legislative and oversight responsibilities relating to the DTIB, an argument also advanced earlier by the Packard Commission.[6] The Secretary wrote that "profound management problems and waste" result from the "redundant phases of budgeting, authorizing, and appropriating defense resources year by year"[7] and that DoD managers are often unable to take needed actions while waiting for uncertain budget authority. He also noted that the large number of congressional committees, subcommittees, and panels with jurisdiction over DoD activities produces "policy gridlock." He wrote that the complexity and lack of coordination in the congressional defense process increases program costs by more than half a billion dollars and causes instability in planning.[8] (See figures 6-2 and 6-3.)

Similar concerns have been echoed by many thoughtful observers. The problem is not only inefficiency but loss of responsibility. A recent book, for example, reported that the average "defense R&D program is voted on by Congress alone an average of 18 times a year in its 8-year life—a total of 144 opportunities to change something."[9] One result of this process is that:

> . . . there is no one individual who feels accountable for results—and when things go wrong there is no one to stand and accept responsibility; there are always lots of persons who can be pointed at as having had their fingers in the pie. The problem is a management structure that permits no single individual to be truly responsible for anything, even at the highest organizational levels—up to and including the President of the United States.[10]

Figure 6-2—Congressional Line Item Changes to DoD Budget Requests

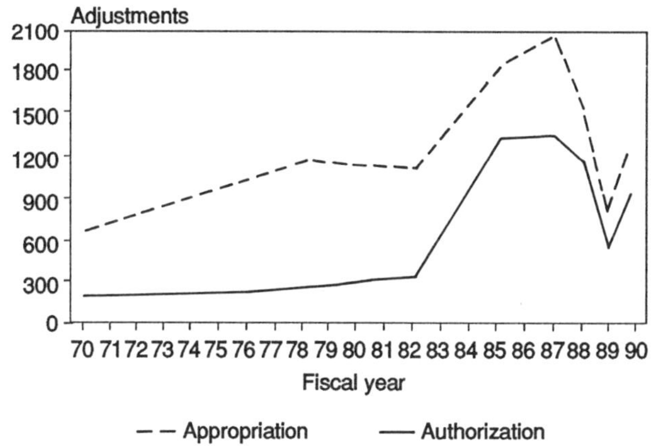

SOURCE: Secretary of Defense, "White Paper on the Department of Defense and the Congress," January 1990.

Such lack of accountability, coupled with the costs of the budget process, have prompted many members of Congress to argue for change. For example, Senator Sam Nunn, Chairman of the Senate Armed Services Committee, has observed that "We are spending most of our time looking at grains of sand on the beach, and we are not looking at the ocean or looking at the horizon."[11]

Congress spends time looking at the "grains of sand" because these represent smaller, potentially solvable problems in which members have an immediate interest; whereas the ocean presents problems that often seem too big and diffuse for Congress to deal with. In addition, Congress steps in wherever it perceives that the DoD has not performed responsibly. The ideal might be for the President to describe an overall strategy, Congress to decide on an overall level of effort, and the DoD to work out details and execute the policy. The reality is that strategies have been difficult to articulate, Congress has had a hard time reaching consensus on the proper level of effort, and the DoD is not just the

[6] The Packard Commission reported:

> Where national defense is concerned, today's congressional authorization and appropriation processes have become mired in jurisdictional disputes, leading to overlapping review of thousands of line items within the defense budget. A growing rivalry between the Armed Services Committees and the Defense Appropriations Subcommittees over the line-item markup of the defense budget has played a major role in moving congressional review of the defense budget toward narrowly focused financial action on individual items and away from oversight based on operational concepts and military effectiveness.

[7] Cheney, op. cit., footnote 3, p. 26.

[8] Report to the President by the Secretary of Defense, *White Paper on the Department of Defense and Congress*, January 1990, p. 1.

[9] Kenneth L. Adelman and Norman R. Augustine, *The Defense Revolution: Strategy for the Brave New World* (San Francisco, CA: Institute for Contemporary Studies, 1990), p. 171.

[10] Ibid.

[11] Steven V. Roberts, "Billions for Defense, The Spending Debate," *The New York Times*, May 17, 1988.

Figure 6-3—DoD Reporting Requirements in Annual Legislation

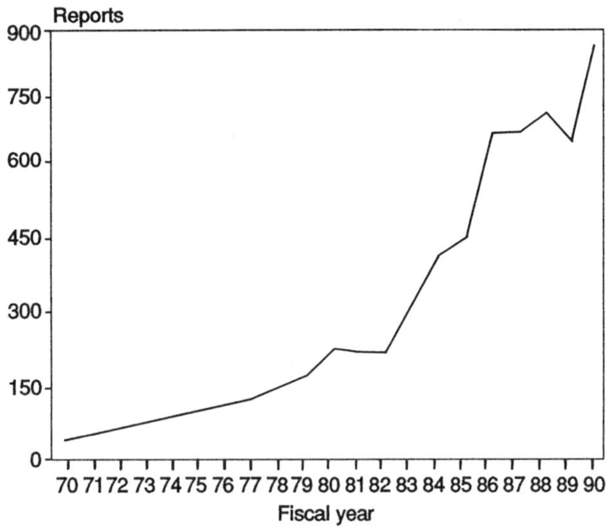

SOURCE: Secretary of Defense, "White Paper on the Department of Defense and the Congress," January 1990.

manager of defense acquisition, but is also, with allies in Congress, the advocate of individual programs. The program-oriented approach has made it difficult to see or organize the big picture.

Once Congress assumes detailed management and control, the DoD loses its incentive to make internal tradeoffs between one weapon and another to accomplish a given mission, since money saved by canceling one program is unlikely to be moved to the other.

To promote efficiency, Congress might give up some of its detailed regulatory role by expanding the DoD's incentives to regulate itself. For example, Congress could reduce the proportion of funds directed at specific programs and expand the proportion allocated to particular missions. When money is allocated to "air defense" rather than to a particular weapon system, then the DoD has a greater incentive to stop work on a troubled system in order to have more resources for a more promising approach to the air-defense mission. The higher the level at which funds are aggregated, the greater are the number of options for achieving a military mission, but Congress' control declines. For example, if the Army is allocated money for air defense, it might choose between guns and missiles, but it is unlikely to consider fighter aircraft. If the allocation is made at the DoD level, the tradeoffs between ground-based and airborne systems can occur. However, by funding at this level, Congress has less influence on which approach is taken.

The danger in making more general funding allocations is that DoD management may produce solutions that a majority in Congress believe are wrong-headed. Congress must choose between giving the DoD responsibility for making choices in the interests of greater efficiency overall and accepting those choices in all but the rarest cases, or maintaining a closer watch over DoD management to avoid the occasional fiasco, while reducing the DoD's incentives to set priorities itself.

Planning

Good, integrated management also requires a defense industry strategic-planning capability that anticipates future national security requirements and considers DTIB alternatives for meeting them. (See box 6-A.)

Planning was addressed in broad terms in the Packard Commission report and more recently in the 1988 DoD report, *Bolstering Defense Industrial Competitiveness*. The DoD report outlined some specific steps, including establishment of a task force to develop a policy on defense industrial planning in support of military operational plans.[12] That task force was never established. But as a part of the DMR, Secretary Cheney directed the establishment of a Defense Planning and Resources Board (DPRB), under the direction of the Deputy Secretary of Defense, to replace the then operating Defense Resources Board. The mission of the DPRB is to "help to develop stronger links between our national policies and the resources allocated to specific programs and forces."[13]

In the absence of DoD actions, Congress has shown considerable recent interest in DTIB planning. It has, for example, mandated an annual plan for developing the technologies considered most critical to ensuring the long-term qualitative superi-

[12] Report to the Secretary of Defense by the Under Secretary of Defense (Acquisition), *Bolstering Defense Industrial Competitiveness*, July 1988. The report recommended, for example, that the DoD "develop industrial strategic plans explicitly linked to military operational plans . . . [and] . . . provide for a continuing assessment of both short- and long-term defense industrial base capabilities," pp. 42-43.

[13] *Defense Management Report to the President*, op. cit., footnote 3, p. 5.

Box 6-A—DTIB Planning Structure

The National Security Act of 1947 addressed the planning requirement squarely. The Act created three boards with planning responsibility for the use of science and industry to support the national security establishment: 1) a National Security Resources Board (NSRB) reporting directly to the President and responsible for formulating policy and plans for industrial and civilian mobilization; 2) a Munitions Board, located in the DoD and composed of Assistant Secretaries of the Army, Navy, and Air Force responsible for planning and coordinating industrial mobilization, production and procurement for the military Services; and 3) a Research and Development Board, also located in the DoD and tasked with developing an integrated R&D program and advising on scientific trends with national-security implications.[1] These organizations never achieved their purpose. The Munitions and Research and Development Boards were abolished in 1953 and their functions transferred directly to the Office of the Secretary of Defense. DoD resource-allocation and major-program planning were later subsumed in the Planning, Programming and Budgeting System (PPBS) established by Secretary of Defense Robert McNamara. The NSRB underwent a series of mergers and reorganizations and ultimately became the current Federal Emergency Management Agency.

[1] See Steven Rearden, *History of the Office of the Secretary of Defense: The Formative Years, 1947-1950*, Historical Office, OSD (Washington, DC: The Government Printing Office), pp. 24-27.

ority of U.S. weapon systems (the DoD Critical Technologies Plan),[14] an annual report on actions taken to improve the ability of the U.S. DTIB to meet national security requirements and the effects on the defense industrial base of defense budgets and plans,[15] and a National Defense Manufacturing Technology Plan.[16] Congress has also strongly urged the establishment of a defense industrial base office within the DoD for the development of policies and plans for the DTIB.[17]

Although such steps appear even more important in the face of a changing international security environment, strategic planning for the DTIB remains controversial. While almost all observers acknowledge the need to ensure that future U.S. military forces have the best scientific and industrial support, some see in DTIB planning the potential for a national industrial policy, which they argue is incompatible with the U.S. free-enterprise system. Others argue that planning is not only essential to match military operations with available resources, but also because the defense industrial base does not operate in a free-market environment. According to this view, the single-buyer relationship between the DoD and defense firms puts a special responsibility on the DoD to plan activities to assure the future health and strength of the DTIB.[18]

Future DTIB planning requires a consensus on what U.S. defense policy will be, what size and types of forces the Nation needs, and what missions they should perform. This consensus has not yet emerged.

Coordination and Cooperation

Integrated management requires coordination and cooperation between the government and the private sector and between the DoD and Congress. The 1988 DoD report *Bolstering Defense Industrial Competitiveness* noted that there are now "deeply ingrained adversarial relationships between Government and industry" and argued that these adversarial relationships "undermine industrial efficiency, responsiveness, and technological innovation."[19] The DMR also noted that the relationship must change, but interviews conducted by OTA with business executives and government officials indicate that a great deal of friction remains. The contracting process is by nature somewhat adversarial and will remain so. Further, the government has a responsibility to ensure accountability of public funds. The tension has been fueled by continued DoD and congressional concerns over unethical behavior by a few defense contractors, combined with intrusive laws

[14] 10 U.S. Code Section 2508.

[15] 10 U.S. Code Section 2509.

[16] 10 U.S. Code, Section 2513.

[17] 10 U.S. Code Section 2503. The office that currently fulfills this function is the Office of Deputy Assistant Secretary of Defense (Production Resources), which includes offices overseeing industrial engineering and quality, manufacturing modernization, commercial acquisition, and production base assessment.

[18] Adelman and Augustine, op. cit., footnote 9, pp. 128-130.

[19] *Bolstering Defense Industrial Competitiveness*, op. cit., footnote 12.

and regulations that often apply criminal sanctions to what might be honest mistakes. Requirements for industry executives to certify, under threat of criminal action, that their firms have properly accomplished numerous activities (many involving paperwork rather than actual production) also costs money and slows the weapons-acquisition process.[20]

The adversarial relationship is not just between the government and industry. The 1990 DoD White Paper on the Department of Defense and Congress noted, for example, that "a final, critical factor affecting congressional defense oversight is a profound lack of trust" flowing from Congress' doubts concerning the competence of DoD managers and the Department's willingness to comply with congressional guidance.[21] Improved intra-government relations will be as important for the future DTIB as will improved relations between government and industry.

Clear Laws and Regulations

Finally, good, integrated management of the DTIB will require clear laws and regulations to guide DTIB activities. Both Congress and the DoD have recognized the need for simplifying the laws and regulations governing resource management and defense acquisition. The DoD Advisory Panel on Streamlining and Codifying Acquisition Laws, mandated by Congress, is a step toward achieving this objective. Simplification is essential to any movement toward increased civil-military integration. Ideally, DoD regulations would be no more onerous than those of the many other government agencies, such as the Departments of Transportation, State, Labor, Commerce, and Justice; the Environmental Protection Agency; the Occupational Safety and Health Administration; and the Equal Employment Opportunity Commission.

There are also many congressional committees other than the Armed Services Committees involved in defense procurement, including Appropriations; Banking; Education and Labor; Energy and Commerce; Foreign Affairs; Government Operations; Intelligence; Science, Space and Technology; Small Business; and Ways and Means.

All of these agencies and committees have different interests, and efficient defense production may not be their number one priority. According to one industry executive, one of the consequences of this current fragmentation of oversight authority is that

> There is no central clearing house for policy, regulations, and oversight nor an integration function in either Congress or the Executive Branch. Yet such an organizational responsibility appears necessary for coordination of any policy recommendations. Without an identified 'change agent' in the U.S. government, it is very difficult for industry to influence the multitude of issues that impact or could impact the industrial base.[22]

In the absence of action to change the situation, problems with "fragmentation of oversight authority" are likely to be compounded in the future, and the laws and regulations that govern defense business are unlikely to be simplified.

The elements of good, integrated management outlined in this section are not particularly controversial in theory but they are extremely hard to implement. For example, many of the Packard Commission recommendations have been accepted only slowly, and even when written into law (e.g., the establishment of a USD(A)), have taken years to have any effect.

There is a fundamental problem of balance between efficiency and accountability in the base. DTIB management should aim to be efficient while accounting for the use of public funds. The need for accountability increases the size of staffs and the numbers of reports. It results in laws and regulations that are more intrusive than they would be if they aimed only at efficient production. Such regulation carries costs that must be weighed against the potential benefits of reducing losses due to fraud, waste, and abuse. A 1989 OTA report illustrated the trade-off graphically (See figure 6-4.) and stated that "Analyses of defense procurement consistently

[20] One firm reported, for example, that it had notified the DoD it would stop supplying a product that previously had been considered a commercial item and supplied at a fixed commercial price. The product's commercial status had changed because it had become obsolete in the commercial world but was still used in the military. Selling the product to the government required adapting to numerous cost accounting requirements and corporate certification of activities with the potential for criminal sanctions.

[21] *White Paper on the Department of Defense and the Congress*, op. cit., footnote 8, p. 20.

[22] Mr. Gordon R. England, executive vice president-Aircraft Programs, general manager-Fort Worth Division General Dynamics, *Statement before the House Armed Services Committee Structure of U.S. Defense Industrial Base Panel*, Oklahoma City, Nov. 1, 1991.

Figure 6-4—Cost v. Regulatory Intensity

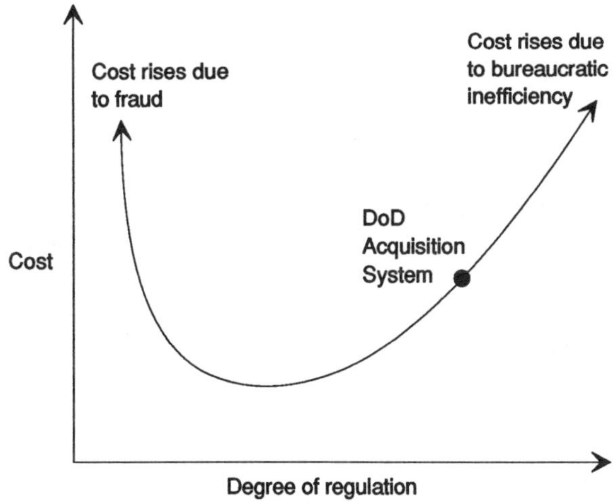

SOURCE: Office of Technology Assessment, 1989.

indicate that the [current] system lies somewhere on the side of excessive regulation, at least in terms of strictly economic consideration."[23] That judgment still appears correct. But the OTA report added that it is possible that "the American taxpayer prefers to pay the high costs of overregulation rather than permit even lesser amounts of public money to go unearned into someone's pocket."[24] Congress will want to consider whether the current amount of regulation is optimal for the new defense era.

ALTERNATIVES FOR MANAGING CHANGE

The immediate challenge facing DTIB management is that the defense budget will shrink markedly in the decade ahead. Earlier chapters argued that a proportional reduction in all sectors of the DTIB would not produce a strong and healthy base. Thus, the United States will face some difficult choices in the coming years.

Redesigning Defense suggested some criteria for making these choices by listing the desirable characteristics of the future base. (See table 1-1.) These criteria ultimately affect specific policy choices, as outlined in chapters 2 through 5. The Nation may, for example, choose to invest relatively more in defense R&D at the expense of production, and close some production facilities and radically restructure those that remain. The Nation may also decide to move more of its R&D capability to the private sector. If so, then some Service research facilities may have to be closed, an action which will require both political and management skills. Similarly, the pursuit of a prototyping-plus strategy, like that outlined in chapter 3, might be accompanied by the reduced production of new weapon platforms, again requiring the closure of some facilities. Three strategies for the transition to a smaller DTIB are discussed below.

A Free-Market Strategy

One alternative advocated for managing the reductions in the DTIB is "allowing the market to decide" which defense contractors will survive in the future. Most of the larger firms that responded to OTA's defense-industry survey favored this approach.[25] These firms argued that the future survivors in the U.S. defense industry should be decided on the basis of which firms win individual contracts. The DoD, in its recent industrial base report to Congress, stated that "in broad context, free market forces will guide the industrial base of tomorrow."[26]

A pure free-market approach would make awards based on the ability to meet each individual contract without consideration of the long-term health of the DTIB. Such an approach would have to end the current practice of using one activity to subsidize another. For example, many observers note that the government often tries to mask the true costs of R&D by having companies support R&D in the expectation of recouping their investment from later production profits.

In the extreme, a free-market approach would allow companies to invest in new products, which they would hope to sell to the DoD at a price that covered R&D costs and adequate compensation for the capital put at risk. This approach could work for a variety of so-called "nondevelopmental items," especially those using the larger pool of commercial technology and production processes. With attention

[23] U.S. Congress, Office of Technology Assessment, *Holding the Edge: Maintaining the Defense Technology Base*, OTA-ISC-420 (Washington, DC: U.S. Government Printing Office, April 1989), p. 152.

[24] Ibid.

[25] See U.S. Congress, Office of Technology Assessment, *Redesigning Defense: Planning the Transition to the Future U.S. Defense Industrial Base*, OTA-ISC-500 (Washington, DC: U.S. Government Printing Office, July 1991), p. 43.

[26] *Report to Congress on the Defense Industrial Base*, op. cit., footnote 4, p. ES-7.

Photo credit: U.S. Navy, 1983

Submarine manufacturing technology may be too specialized to be supported by commercial industry alone.

focused on the comparison of the price and value of a product, there would be no need for internal accounting to establish production costs, since the price to the government, not the cost to the company, is the only financial concern of the government in the role of free-market customer.

Industry self-financing would work less well for specialized military products or entire weapon platforms. In these cases, defense R&D would almost certainly have to be funded by the government. Whichever firm was judged best qualified to carry out the development would win the development contract. Afterwards, production contracts could be let using the same criterion. In those sectors where specialized production houses (so-called "build-to-print" shops) consistently won the production contracts, the free market would evolve toward a division of labor between design houses and production houses. One criterion for awarding a development contract could be that the product be producible in the largest possible number of commercial facilities. In many sectors, however, some advantage would accrue from having the R&D function and production under the same roof. In these sectors, build-to-print shops would not be able to compete, and more integrated firms would evolve.

Multiple sources would survive only when such a market structure was more efficient than one composed of single sources; a pure free-market approach would not admit to giving contracts to second-place finishers just to maintain alternative sources of supply.

Many observers, as noted elsewhere in this report, argue that pure free-market mechanisms are impossible to apply because the defense industry simply does not operate in a free market, but is instead a regulated monopsony. But a number of these observers argue that eliminating many of the current legislative and regulatory restrictions on defense acquisition will open the defense market to increased competition. This increased civil-military integration of the base may promote the use of a free-market approach in many sectors of the future DTIB.

An Activist Strategy

A second management alternative is a more activist approach, stressing government participation in implementing the changes in the DTIB. The activist approach includes a range of proposals. Some advocates want to select surviving defense firms and support the development of defense technologies having civilian application. The argument for such support is that the broader national technology and industrial base is essential to future U.S. military strength.[27] Advocates of an activist approach see little potential for free-market operation in most of the defense sectors, given the DoD's role as single buyer. As a result, while they support changing the regulatory environment to permit the use of commercial practices, they also favor more government intervention to enhance specific technologies and industrial sectors.

An activist approach is used in France, for example, where government planners allocate defense work to ensure the competitiveness and financial health of the French defense industry. To preserve a key design team, for example, the French procurement agency, the General Delegation for Armaments (DGA) may award a development contract on a competitive basis but give the loser a share of the subsequent production work to keep both firms in business, even if total procurement costs are thereby increased. Similarly, the DGA may procure a system from a French firm even when it could be acquired faster and more cheaply from a foreign source; and it may keep an assembly line

[27] The DoD report *Bolstering Defense Industrial Competitiveness*, for example, examined the negative trends in such dual-use industrial sectors as machine tools and electronics and sought to develop policies to help change these trends.

open between procurement cycles by stretching out the rate of production until the next contract comes along.[28] This approach has sometimes involved making a choice between buying weapons that the French armed services desired or buying weapons to support elements of the French defense base.

A Mixed Strategy

The successful management of the future, smaller DTIB will probably involve elements of both the free-market and activist strategies. Where sufficient real competition exists, the free-market approach of providing funds to the successful bidder will be satisfactory. But this competitive environment is limited to particular technologies, mostly subtier industries making components common both to military systems and civilian products. In other areas, where a defense technology or industrial sector has little or no civilian counterpart, source selection on a nonmarket basis is more appropriate.

The DoD appears, in practice, to support this mixed strategy approach. While the DoD has been criticized for placing too much reliance on market forces in defense procurement, the Department acknowledges that the free market alone might not provide the necessary industrial capability in selected areas. For example, because the U.S. shipbuilding industry is dependent on Navy business, the DoD has stated that it "will require continuous monitoring by the DoD to ensure a capable prime contractor and supplier base is available for military needs."[29] Secretary Cheney has also said that his decision to build another nuclear-powered aircraft carrier was based, in part, on the need to preserve the Navy's shipbuilding industry.[30] The DoD industrial base report suggests the possibility for intervention in the armored-vehicle sector, although the Department has not yet taken any action to preserve production capabilities in this area.

How much government intervention is necessary will depend on how the Nation structures the future DTIB. Those advocating greater civil-military integration argue that integration will strengthen the free market and ultimately reduce the need for government intervention. **But civil-military integration will occur only at the price of modifying some of the regulatory mechanisms currently built into the procurement system to ensure public accountability of funds.**

Any strategy to restructure the DTIB requires the government to have a clear vision of the future defense establishment and the DTIB needed to support it. That vision must be communicated to industry. In the words of one defense contractor:

> It is our view that the White House, OSD, and Congress must articulate and agree on a national defense policy to avoid the chaos of a tear-down rather than an orderly build-down. If a tear-down occurs, the quality of defense products is likely to suffer badly.[31]

FUTURE MANAGEMENT TOOLS

Beyond the immediate problems associated with down-sizing, there are longer range concerns over how to manage the future base to get the most return from a smaller and much changed DTIB. Steps that might be taken to improve the management of the future base are outlined below.

Improved Planning

Future DTIB planning must be better coordinated. Today, DTIB planning remains relatively decentralized within the DoD. The individual Services develop plans that are further subdivided into R&D plans, production plans, and depot maintenance plans. Decentralized planning has the benefit of staying close to Service requirements. Yet if it results in costly redundancies and bottlenecks in industrial responsiveness, it will fail to meet either the immediate military needs of the commanders-in-chief of the unified and specified military commands or the longer term needs of the DTIB. Better coordination among the three elements of the base (R&D, production, and maintenance) will help reduce the past tendency of DTIB managers to make decisions that seem best for their organization, but actually have negative implications for the base as a whole.

[28] U.S. Congress, Office of Technology Assessment, *Lessons in Restructuring Defense Industry: The French Experience*, OTA-BP-ISC-96 (Washington, DC: U.S. Government Printing Office), June 1992.

[29] *Report to Congress on the Defense Industrial Base*, op. cit., footnote 4, p. ES-5.

[30] "Industry Base, Fee Structure Entered Carrier Decision—Cheney," *Defense Daily*, Feb. 6, 1992, p. 200.

[31] Response to OTA industrial base survey.

The growing importance of the broader national and global technology and industrial bases in meeting defense requirements increases the need to bring other government agencies into DTIB planning. The Federal Emergency Management Agency (FEMA) is responsible for emergency and mobilization planning involving the civilian agencies. The Department of Commerce might share more peacetime DTIB planning responsibilities with the DoD. Industry must also be more directly involved in the planning process. FEMA has taken a number of initiatives to increase the understanding of the roles that the civil agencies must play in supporting the DTIB.

One of the chief criticisms of current planning is the lack of good information on the DTIB. Operating in a more integrated base with fewer resources will require a better understanding of not only the DTIB, but also the larger national base. The Critical Technologies Plan and the Industrial Base Report appear to have been partly motivated by a desire to have the DoD collect better data and thus develop a better understanding of the base. The DoD has supported several industrial base data-gathering efforts, but has never placed high priority on them.

Redesigning Defense examined a number of current government and industrial databases and concluded that all "are short of data because data collection efforts are generally underfunded and are not standardized."[32] For example, the Defense Industrial Network (DINET) sought to link a number of commercial and DoD industrial databases in order to provide insight into the condition of the DTIB. But according to many observers, this effort was underfunded. In other cases, support for setting up a database has not been followed up with adequate funding for data collection to make that base useful. Systematic data gathering is expensive.

The DoD should set priorities on what data to gather and how much data are needed to manage the base. But, there are other agencies (e.g., such as Department of Commerce and FEMA) that should play a more active role. **Congress can act to ensure that industrial-base data priorities are established by the Executive Branch and that adequate funding is available.**

Figure 6-5—Integration of the DTIB Elements

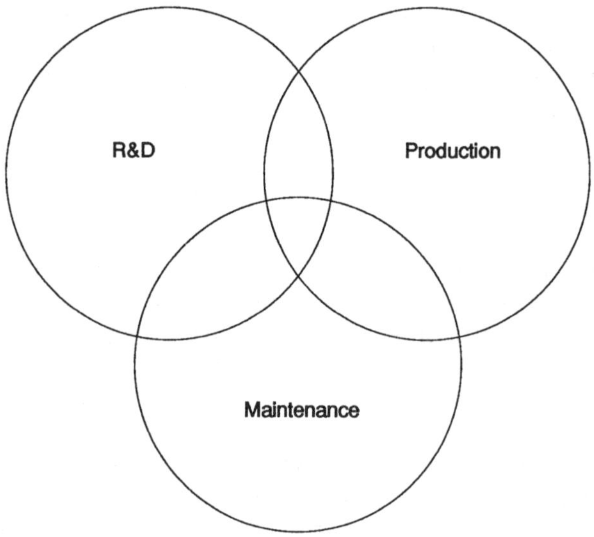

SOURCE: Office of Technology Assessment, 1992.

Organizational Changes

The future DTIB must have more integration of the R&D, production, and maintenance elements. (See figure 6-5.) The Air Force appears to have begun this process by combining its Air Force Systems Command and Air Force Logistics Command into a single Air Force Materiel Command (effective July 1, 1992). The other Services are also combining elements of their commands.

But more important than the major command reorganizations will be changes involving programs and technologies. The Air Force's new concept of integrated weapon system management (IWSM) is currently being tested on 21 weapons programs.[33] The IWSM establishes a program director in charge of all aspects of the life cycle of a weapon system from R&D through production and maintenance. The program office is located in a product division during program development, but moves to a logistics center once the system has been produced and is operational.[34]

While a concept such as the IWSM integrates the management of the three principal elements of the DTIB, it may not be adequate in the future environ-

[32] *Redesigning Defense*, op. cit., footnote 25, p. 116.

[33] John Terino, "Doing Business," *National Defense*, January 1992, pp. 18-19.

[34] United States Air Force Fact Sheet, *Integrated Weapons System Management: Cornerstone of Air Force Materiel Command*, Nov. 1, 1991.

ment because it is focused on managing weapon systems. It does not break the present near-automatic link between development and production. **In the future, research, development, and prototyping will be pursued without the expectation that production of a final system will necessarily follow. Therefore, DTIB management cannot be centered around individual weapons programs.** A useful additional concept is that of integrated mission area management. This approach would examine alternative ways of achieving a mission as well as the tradeoffs among R&D, prototyping, production, and maintenance in sustaining an overall DTIB capability to support identified national security requirements. This is now done by the Services in their mission area analyses and in their management of technologies (e.g., the Tank Automative Command looks across the armored vehicle sector). But current efforts are largely limited to programs within a single Service. The Joint Staff might assume the job of analyzing missions among the Services.

Degree of Centralization

The degree of centralization of future DTIB management is an important issue. Secretaries of Defense have generally pushed for increased control of resources and acquisition (Secretaries Laird and Weinberger were exceptions), while the Services have sought more autonomy. Proponents of more centralized acquisition argue that current inter-Service coordination will be insufficient to manage the future DTIB so that it will be suitably strong and flexible. Proponents of decentralization counter that centralization will separate equipment acquisition from the military users.

Several forms of centralization have been proposed. Three possibilities are:

1. a "purple suit" (i.e., joint-Service) procurement agency that would buy all military hardware and supplies;
2. a division of procurement tasks so that each Service is responsible for supplying the others with a set of procurement items; or
3. an independent acquisition corps separate from the Services, staffed by civilians.

A joint-Service agency might resemble an expanded Defense Logistics Agency, which is currently responsible for providing common items, such as fasteners, food, and uniforms.[35] This new agency would take advantage of the long-term experience of career civilian procurement officers and the military expertise of Service officers.[36] Its facilities would need to be geographically close to Service technical centers.

Alternatively, tasks could be distributed according to Service expertise or priority. Thus, the Air Force might be responsible for all cargo planes, the Army for all trucks, the Navy for all boats and ships, and the Marines for landing craft. The DoD has run joint procurement programs; some successful (e.g., trucks and 20mm ammunition), and others less successful (e.g., F-111).[37]

A separate civilian acquisition corps could break direct Service advocacy for developed systems to enter production. But a drawback of this approach is that an independent organization can easily lose sight of Service requirements. The French acquisition agency, for example, has been criticized for not being responsive enough to battlefield requirements.[38]

The goals of any reforms should be to reduce redundancy, to make larger, more economical purchases, and to have an experienced cadre of acquisition personnel who do not have a direct Service stake

[35] The Defense Logistics Agency also procures replacements for microcircuits that have gone out of commercial production but are still required for maintaining a weapon system. Many microcircuits have a product cycle of about 7 years, while the systems they are a part of may have an operational lifespan of 20 years or more. See Donald O'Brien, testimony before the U.S. Congress, House Armed Services Panel on Future Uses of Manufacturing and Technology Resources, Oct. 24, 1991.

[36] The French have a centralized procurement agency, which has resulted in multi-Service procurements and better coordination of R&D investment. Nevertheless, critics allege that the agency has overemphasized industrial base considerations at the expense of military requirements and force readiness. See U.S. Congress, Office of Technology Assessment, op. cit., footnote 28.

[37] The F-111, originally designated the TFX (tactical fighter, experimental), began as a biservice program run by the Air Force to meet Navy fleet air defense and Air Force deep strike requirements. The differences between the two variants of the plane were to be minimal and the use of common parts was to be emphasized. Secretary of Defense McNamara believed that joint procurement and commonality would save the Nation about $1 billion. In 1968, 7 years after the program began, Congress canceled the naval variant of the F-111, ostensibly because the effort to achieve commonality undermined the planes' ability to carry out their different missions and was not cost effective. See U.S. Senate, Government Operations Committee, Permanent Subcommittee on Investigations, *Hearings on the TFX Contract Investigation (Second Series),* Part I, Serial No. 43-096, Mar. 24, 1970.

[38] *Lessons in Restructing Defense Industry* , op. cit., footnote 28.

Photo credit: U.S. Department of Defense

Congress canceled the Navy's F-111 variant after an experiment in joint procurement and commonality with the Air Force failed.

in getting a particular weapons program into production.

Congress may also wish to reconsider its committee structure for overseeing the DTIB. Oversight of R&D, acquisition, and maintenance could be consolidated. Congress may wish to consider reducing the number of committees, subcommittees, and panels responsible for DTIB issues just as it calls on the DoD to be less top-heavy in its management of the base.

Degree of Civil-Military Integration

The Federal Government's role as single buyer in the military market gives it enormous power to shape that market. In the past, the DoD has been such a large customer that it could establish unique and sometimes onerous requirements—in accounting, manufacturing, and management—and still be confident that sellers would step forward to seek its business. This heavy regulation has isolated the DTIB from the broader national base.

The burden of regulation was not financially crippling as long as the DoD market remained large enough on its own to support entire companies. But, excessive regulation will become a major obstacle to maintaining a healthy DTIB as the defense market shrinks and becomes less attractive to private firms.

Integrating the defense base back into the larger industrial base will require changes. Reducing the amount of government oversight of firms' defense work is possible—but only in response to assurances that there is proper corporate accountability. Two general courses might be followed. One involves changes within the current defense acquisition system. Examples include such programs as the Corporate Risk Assessment Guide (CRAG) developed by the Defense Contracts Auditing Agency to reduce the number of on-site inspectors in key financial areas; the Exemplary Facility (EF) program, which has been tested in a number of manufacturing facilities in the past 2 years; and the Army's Continuous Process Improvement Program. While such programs have the objective of reducing oversight and therefore reducing costs, they all suffer from inadequate government support—especially a lack of support by relevant DoD oversight agencies—and a subsequent lack of industry incentive to participate. For example, the EF program was recently discontinued by OSD with little discussion with the companies involved. Future efforts to reform the acquisition system will require broad-based support within the DoD if they are to succeed.

The second course is to make much wider use of "commercial standards" in auditing and production, i.e., a broad, direct effort at increased civil-military integration of the base. This course offers greater potential benefits than limited change within the DoD system. For example, acceptance of commercial standards in place of military standards (e.g., replacing Mil-Q-9858A with ISO 9000) has been proposed by many in industry, but has not been acted on favorably by the DoD.[39] (See figure 6-6.) Even if this change were made, the DoD's need for accountability would be different from that of the civil sector. Advocates of civil-military integration argue that, nonetheless, the regulatory barriers to doing DoD work should be lowered and more firms brought into the defense business, at which point accountability can be better assured through real competition.

Much of the burden of government accounting, auditing, and management regulations derives from

[39] ISO 9000 is shorthand for International Standards Organization (ISO) 9000- 9004, a series of documents on quality assurance published by the Geneva-based ISO. The 5 documents outline standards for developing Total Quality Management and a Quality Improvement Process. 9000 consists of guidelines for the selection and use of the quality systems contained in 9001-03. 9001 outlines a model for quality assurance in design, development, production, installation, and servicing. 9002 outlines a model for quality assurance in production and installation. 9003 outlines a model for quality assurance for final inspection and testing. 9004 is not a standard but contains guidelines for quality management and quality system elements.

Figure 6-6—U.S. Standards

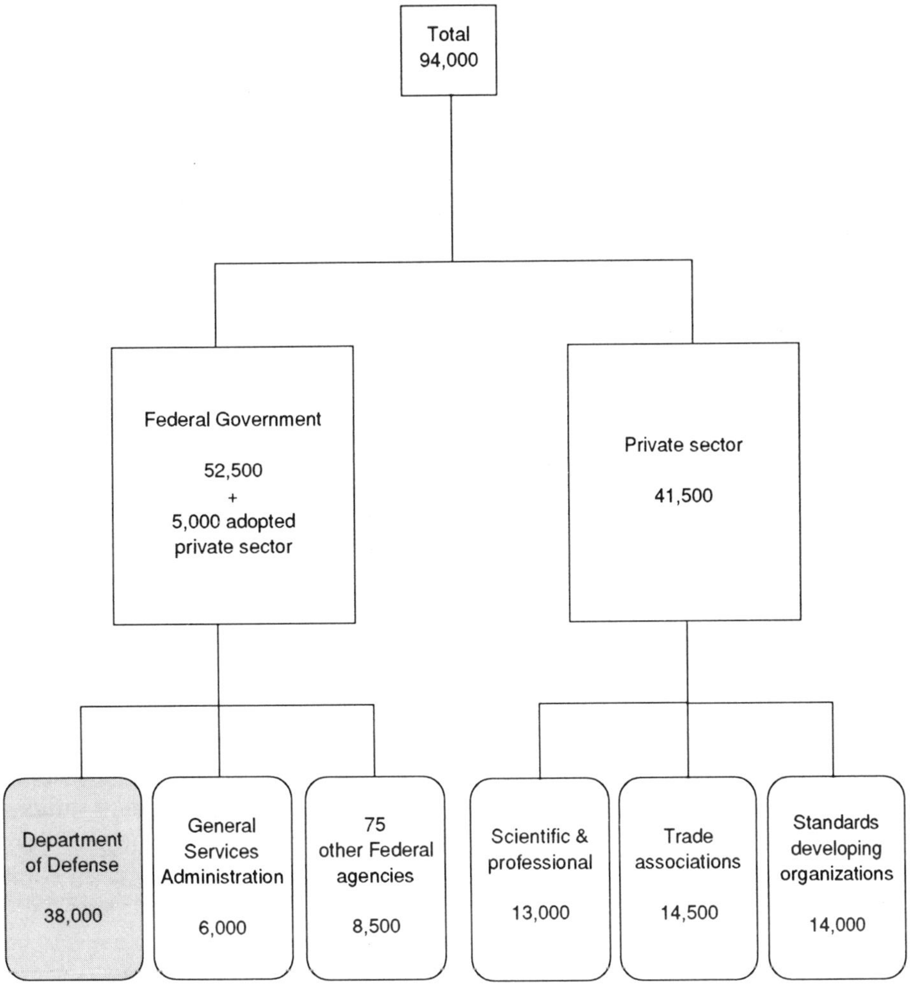

SOURCE: Robert Toth, Toth Associates.

a desire to account for the actual production costs of weapons rather than the final price paid by the government, and to identify and punish cases of fraud, waste, and abuse. In principle, once a fair price has been established, the government should be satisfied with that price for future identical items. In practice, accounting oversight continues. Further, the laws and regulations governing accounting emphasize the potential for criminal sanctions to be imposed for what many in industry view as simple mistakes; these rules ultimately raise costs and encumber the acquisition process. In the past, Congress has been a major force behind these laws and regulations and may now wish to reconsider them.

Advocates of increased reliance on civilian business practices argue that market incentives would help control costs and maintain accountability. Manufacturers would be evaluated on their ability to produce items of agreed quality on time and at, or below, contract price. Military specifications would focus on desired performance and less on detailed manufacturing procedures. It might be important to continue specifying critical procedures (e.g., specialized welding), but such process specifications might be made advisory in many cases.[40] Defense contractors could be given more flexibility and responsibility in carrying out their obligations, allowing more efficient integration of commercial and defense facilities, equipment, and supplies,

[40] Caution must be taken to ensure that "advisory" specifications do not become *de facto* specifications.

which are now largely segregated, and reducing overhead spent on paperwork. The DoD has begun to test the effectiveness of commercial practices in lowering costs while guarding against abuse. It could widen such practices in the future.

The current Advisory Panel on Streamlining and Codifying Acquisition Laws is scheduled to report in January 1993. It is expected to provide insight into how Congress might best reform the vast body of acquisition law that was built up during the cold war.

Improving Cooperation and Coordination

Combined government and industry action will be critical in managing the transition to a production base with the desired characteristics discussed in chapter 4. Unfortunately, as noted earlier, government-industry relations are often adversarial. Improved cooperation and coordination is needed not only between industry and the DoD, but also between the DoD and Congress and between Congress and industry.

This improved cooperation could begin in the planning phase. Industry needs to understand future defense requirements so it can prepare to meet them. To achieve such coordination, attitudes on all sides will have to change. Deputy Defense Secretary Donald Atwood has acknowledged the need for more cooperation and more industry participation in defense planning, noting that in the past it was considered "a crime if you [business] knew what we're going to do."[41] Since long-term DoD planning goals were classified, industry had difficulty preparing for the future. A more open approach is needed so all sides can make coordinated plans in a way that reduces any incentive to cheat.

Talented and Experienced People

While program managers and contracting officers are certainly key players, management of the DTIB involves more than running acquisition programs. **Future DTIB managers will have to make trade-offs among the three principal elements of the base (R&D, production, and maintenance), while ensuring that the desired future DTIB capability will be available.** Dealing with these broad and basic questions demands experienced personnel.

Photo credit: Defense Systems Management College

A lecturer conducts a management course at the Defense Systems Management College.

"Revolving-door" laws make it difficult to attract talented and experienced people as senior civilian DTIB managers. These laws limit the post-government activities of appointees and require that they divest themselves of current stocks or commitments that might be a conflict of interest while they serve the government. Such disincentives to government service could become even greater if a strategy of increased civil-military integration is pursued, since senior managers of non-defense firms might also be dissuaded from DoD service. Thus, while conflict-of-interest laws are essential, they might be reformed with an eye to attracting experienced private-sector managers to DoD jobs.

The quality of the Services' acquisition workforce has also been criticized. The principal problems appear to be rapid turnover among uniformed program managers, inexperience, and inadequate educational backgrounds. The Defense Acquisition Workforce Improvement Act (DAWIA) sought to address some of these problems. The military has set up a Service Acquisition Corps and acquisition career paths. However, since over 90 percent of the personnel in the acquisition workforce are civilians,[42] programs to enhance military personnel are not enough to improve DTIB management. Greater efforts have to be made toward civilian managers.

[41] Deputy Secretary of Defense Donald J. Atwood, speech to a Technical Marketing Society of America Seminar in Arlington VA, reported in *Aerospace Daily*, June 3, 1991, p. 370.

[42] U.S. House of Representatives Committee on Armed Services, Report of the Subcommittee on Investigations, *The Quality and Professionalism of the Acquisition Workforce*, May 8, 1990, p. 14.

One problem is that civilian salaries lag behind military pay.[43]

DAWIA also provides for better training of acquisition personnel. The acquisition university, mandated by the Act is now being established. About 75 students per year will receive specialized acquisition courses at the Industrial College of the Armed Forces. Senior DTIB managers need training in how to maintain a "warm capability" involving tradeoffs among all three elements of the base, and how to manage a base that is more integrated with the civilian sector.

One approach to maintaining a cadre of technical managers, as well as a labor force, has been suggested by William L. Clark of the Defense Systems Management College. Training personnel at all levels could be accomplished by establishing a Civilian Technical Reserve Corps—a volunteer group of skilled defense engineering and production personnel who would take periodic "updating" and retraining in their particular specialties. Some of this person-to-person training might be aimed at younger people to preserve generational continuity.

In case of a national emergency, a cadre of such trained individuals would greatly facilitate DTIB mobilization and, in particular, the transition of prototyped systems to quantity production. A precedent for this kind of continuing education already exists in the medical profession and, to a lesser degree, in the legal profession. Volunteer reserve forces also exist in the military and thc Pcacc Corps. All of these have successfully harnessed personal pride and patriotism to serve well-defined national goals. A Civilian Technical Reserve Corps would require participation by industry management, organized labor, and human resource experts.

SUMMARY

For the Nation's DTIB to remain strong and healthy in the future, it will have to be managed in new ways. There must be greater centralization in planning, more flexibility in operation, and increased integration with the larger civilian technology and industrial base. **A coherent management strategy for the entire base (R&D, production, and maintenance) will be critical to halt the weakening of the present base and ensure that a "build down" rather than a "tear down" occurs.** Centralized strategy will need to be combined with decentralized operation, giving individual managers more responsibility and authority. Managers will have to make tradeoffs with respect to the entire base and not just a single element. DoD managers should have more flexibility in dealing with industry, and DoD contractors should have more authority to deliver products and services in a way that is most efficient for them and the base as a whole. These shifts will require experienced and talented management personnel—who will only be available if changes are made in training, career paths, and pay.

Sweeping reform of the acquisition laws and regulations will be essential to achieve the flexibility needed for restructuring. Congress might support programs that reduce direct government oversight, or change the oversight process radically and move to a more commercial environment. Such reforms should differ somewhat according to the industrial sector. Those defense sectors with more civilian overlap might best be managed according to civilian standards, while militarily-unique technologies will need more specialized management.

Finally, successful DTIB management will require better coordination and cooperation between government and industry and between the executive branch and Congress. Consensus on the role that the DTIB plays in national security is needed. If past inefficiencies are allowed to persist, the much smaller future base will be unable to provide the required support.

[43] Ibid., pp. 480-487.

Appendixes

Appendix A

Defense Technology and Industrial Base Policies of Allied Nations

U.S. strategies for restructuring the defense technology and industrial base (DTIB) will be influenced by the DTIB strategies of our principal allies as well as any potential adversaries. For example, allied emphasis on collaborative procurement may affect the tendency of the United States to engage in such efforts. Also, international arms sales will affect both U.S. sales and possible levels of U.S. R&D. This appendix provides a tabular overview of foreign DTIB structures and policies for Canada, France, Germany, Japan, and the United Kingdom. The defense bases of these nations are all facing pressures similar to those on the U.S. DTIB.

Table A-1 describes the structure of the DTIBs of the five countries. Indicators are the size of the industry (measured by personnel), percent owned by the state, and export sales.

Table A-2 indicates current spending and trends in defense spending (where available), including investment in military R&D and procurement.

Table A-3 lists the countries' goals for restructuring their defense bases, such as what design and manufacturing capabilities each country wishes to preserve, priorities for defense R&D and procurement, and plans for surge production or industrial mobilization in crisis and war.

Finally, table A-4 describes government strategies for achieving the desired goals.

Table A-1—Structure of the Allies' DTIBs

Country	Industry size (direct employees)	Percent state-owned	1990 Export sales (billions of U.S. dollars)
Canada	80,000	0%	$0.88[b]
France	260,000	80%	$6.7
Germany	265,000	almost 0%	NA[c]
Japan	NA	0%	0
United Kingdom	300,000[a]	almost 0%	$5.4

NOTE: Conversion rates: $1 = 0.56£; $1 = 5.5FF; $1 = 1.19 Cdn.
NA = not available.
[a] Financial year 1989/90. Includes 150,000 jobs sustained by defense exports.
[b] Most exports are to the United States.
[c] A relatively small amount.
SOURCE: Office of Technology Assessment, 1992.

Table A-2—Fiscal Year 1992 Allied Defense Spending
(billions of U.S. dollars)

Country	Total defense budget	Defense R&D budget	Procurement budget	Percent GNP
Canada	$10.6	$0.12	$ 5.1	1.8%
France	$35.5	$5.5	$18.7	3.3%
Germany	$40.2	$1.8	$ 4.8	2.5%
Japan	$34.2	$0.93	$ 6.5	0.9%
United Kingdom	$43.1	$4.7	$16.1	4.0%

NOTE: Defense budgets converted to U.S. dollars using the following exchange rates
$1 = 1.19 Cdn
$1 = 135 ¥
$1 = 1.64 DM
$1 = 5.5 FF
$1 = 0.56 £

SOURCE: Office of Technology Assessment, 1992.

Table A-3—Priorities for Restructured DTIBs

Country	Core capabilities	R&D	Procurement	Surge/mobilization
Canada	Complex subsystems, shipbuilding, light armored vehicles, trucks, small arms, helicopters.	Focus at subsystems level; government facilities specialize in militarily unique technologies.	Canadian patrol frigate, light armored vehicles, tactical command and control systems, helicopters.	Institutionalized defense industry planning, stresses cooperation with the United States.
France	Full range of major tactical weapon platforms, nuclear weapons.	New emphasis on space systems, command and control, and guided standoff missiles.	*Rafale* fighter, *Leclerc* tank, *Améthyste* submarine, *Charles de Gaulle* carrier, *Hélios* and *Syracuse* satellites.	Little emphasis on planning by Services.
Germany	Issue is currently under study.	Stresses joint development programs, aerospace.	Major cuts over next decade in heavy armor. Continued commitment to the European Fighter Aircraft in question.	No detailed defense industrial mobilization planning.
Japan	Develop wartime maintenance and supply capabilities, relies on civilian R&D, supports aircraft electronics.	Increased emphasis on defense R&D, aircraft, missiles, logistics and support, but relies on civilian technical developments in key high technology areas.	Present program stresses improvement and modernization of existing equipment.	No detailed defense industrial mobilization planning.
United Kingdom	Determined by market forces.	Aviation, stealth, and electronics.	*Challenger* tank, European Fighter Aircraft, attack helicopter, nuclear deterrent.	Currently relies on limited, ad hoc planning, but may move to more structured planning.

SOURCE: Office of Technology Assessment, 1992.

Table A-4—Allies' DTIB Strategies

Country	National plan	International collaboration	Civil-military integration	Consolidation
Canada	Focus is on continued close cooperation with U.S., limited government intervention, and increased government/industry consultation.	Participates in NATO's Conventional Armaments Planning System, but this is not a major policy thrust. Efforts are encouraged at the firm level.	Recognition of increased importance of dual-use technologies, closer ties between defense and civil R&D organizations.	Relying on market forces, foreign demand.
France	Central government strategy is to maintain areas of excellence in French defense industry, stress international sales.	Systematic approach to European collaboration and strategic alliances, but go-slow approach to free arms market within the EC.	Government encourages diversification of firms, no barriers to civil-military integration.	Government promoting some consolidation, cross-border mergers.
Germany	Free-market orientation, with close exchange of information between government and industry.	Strong and growing emphasis on collaboration.	Stressing civilian products where militarily acceptable.	Industry is down-sizing, government currently sees no need for major additional restructuring of base.
Japan	Limited defense planning, stress on U.S. relationship, use of dual-use technology.	Strictly limited by law, cooperation with U.S. is viewed as important.	Considerable integration; most defense firms produce civilian products, but a few firms produce most Japanese defense items.	Ostensibly left as a corporate decision, but most Japanese industrial-sector decisions involve government administrative guidance.
United Kingdom	Reliance on the private sector, greater civil-military integration, exports, limited government intervention.	Supports collaboration with allies, expects it to increase as budgets are reduced and forces become more international.	Key component—relaxed requirements to permit use of civil technology, most defense firms diversified into civil sector.	Relying on market forces, government provides information to industry about future defense plans and intentions.

SOURCE: Office of Technology Assessment, 1992.

Appendix B
Reviewers

This report has benefited from the advice of many individuals from the government and the private sector. OTA especially would like to thank the following individuals for reviewing various drafts of the report. The views expressed in this report, however, are the sole responsibility of the Office of Technology Assessment.

LTC Diane Beardsley
Headquarters
Air Force Logistics Command
Wright-Patterson AFB, Ohio

Don Carson
TASC
Arlington, VA

William Clark
Defense Systems Management College
Ft. Belvoir, VA

Richard E. Clemens
BMY-Combat Systems
York, PA

James M. Compton
Office of the Secretary of Defense
Department of Defense
Washington, DC

Jim DeZutter
Digital Equipment Corp.
Marlboro, MA

Gordon R. England
General Dynamics Corp.
Fort Worth, TX

S. Bruce Ensley
Department of the Navy
Washington, DC

Don Ferguson
Bell Helicopter Textron, Inc.
Fort Worth, TX

Donald Fowler
F&F Associates, Inc.
Arlington, VA

Gary W. Fredricks
General Dynamics Corp.
San Diego, CA

George L. Friedman
Northrop Corp.
Los Angeles, CA

Lucy W. Fyfe
Pratt & Whitney
West Palm Beach, FL

Steven R. Goldberg
ITT Defense Avionics
Clifton, NJ

Joe P. Golden
Office of the Secretary of Defense
Department of Defense
Washington, DC

Jack S. Gordon
Lockheed Advanced Development Co.
Sun Valley, CA

Jay W. Gould III
Defense Systems Management College
Ft. Belvoir, VA

Col. Robert Hayes
United States Air Force
Washington, DC

Del Jacobs
Northrop Corp.
Hawthorne, CA

Major General Harry G. Karegeannes
U.S. Army Depot System Command
Chambersburg, PA

Chuck Kimzey
Office of the Secretary of Defense
Department of Defense
Washington, DC

Leslie M. Lackman
Rockwell International Corp.
Los Angeles, CA

Arnold Levine
RAND Corp.
Santa Monica, CA

Henry J. Lopez
Allied-Signal Aerospace Co.
Torrance, CA

Robert Mason
Office of the Secretary of Defense
Department of Defense
Washington, DC

Bettie McCarthy
Proprietary Industries Association
Washington, DC

William W. Mogan
Talley Defense Systems
Mesa, AZ

Joseph Moreland
Federal Emergency Management Agency
Washington, DC

Jimmy C. Morgan
Munitions and Chemical Command
Rock Island, IL

Robert Morgan
AIResearch, Los Angeles Division
Torrance, CA

William Motley
Defense Systems Management College
Ft. Belvoir, VA

Joseph Muckerman
Office of the Secretary of Defense
Department of Defense
Washington, DC

Daniel J. O'Neil
The University of Oklahoma
Norman, OK

George A. Paulikas
The Aerospace Corp.
El Segundo, CA

Robert Pendley
Los Alamos National Laboratory
Los Alamos, NM

George E. Pickett, Jr.
Northrop Corp.
Arlington, VA

Arati Prabhakar
DARPA/MTO
Arlington, VA

Winn Price
Bath Iron Works
Bath, ME

Thomas W. Rabaut
FMC Corp.
Santa Clara, CA

Otto Renius
General Dynamics Land Systems
Sterling Heights, MI

Frederick Riddell
Institute for Defense Analyses
Alexandria, VA

Howard Roberts
General Dynamics Land Systems
Sterling Heights, MI

Mike Robinson
Rockwell International Corp.
Los Angeles, CA

Fred E. Saalfeld
Office of Naval Research
Arlington, VA

Philip A. Selwyn
Office of Naval Technology
Arlington, VA

Giles Smith
RAND Corp.
Santa Monica, CA

H.C. Smyth, Jr.
Bell Hellcopter Textron Inc.
Arlington, TX

William M. Tolles
Naval Research Laboratory
Washington, DC

Nicholas M. Torelli, Jr.
Office of the Secretary of Defense
Department of Defense
Washington, DC

Tom Welch
Office of the Secretary of Defense
Department of Defense
Washington, DC

William R. Widmer
General Dynamics Corp.
Tallahassee, FL

Dennis J. Wrightman
Logistics Management Institute
Bethesda, MD

T.G. Yuguchi
General Dynamics Corp.
San Diego, CA

Superintendent of Documents **Publications** Order Form

P3

Order Processing Code:
* **6313**

☐ **YES**, please send me the following:

Charge your order. It's Easy!

To fax your orders (202) 512–2250

_________ copies of ***Building Future Security: Strategies for Restructuring the Defense Technology and Industrial Base (168 pages)*** S/N 052-003-01289-8 at $9.50 each.

The total cost of my order is $_________. International customers please add 25%. Prices include regular domestic postage and handling and are subject to change.

(Company or Personal Name) (Please type or print)

(Additional address/attention line)

(Street address)

(City, State, ZIP Code)

(Daytime phone including area code)

(Purchase Order No.)

May we make your name/address available to other mailers? **YES** ☐ **NO** ☐

Please Choose Method of Payment:

☐ Check Payable to the Superintendent of Documents

☐ GPO Deposit Account ☐☐☐☐☐☐☐ – ☐

☐ VISA or MasterCard Account

☐☐☐☐☐☐☐☐☐☐☐☐☐☐☐☐☐☐☐☐

☐☐☐☐ (Credit card expiration date)

Thank you for your order!

(Authorizing Signature)

6/92

Mail To: New Orders, Superintendent of Documents
P.O. Box 371954, Pittsburgh, PA 15250–7954

ISBN 0-16-037975-X
9 780160 379758
90000